MW01356232

Springer Geography

The Springer Geography series seeks to publish a broad portfolio of scientific books, aiming at researchers, students, and everyone interested in geographical research.

The series includes peer-reviewed monographs, edited volumes, textbooks, and conference proceedings. It covers the major topics in geography and geographical sciences including, but not limited to; Economic Geography, Landscape and Urban Planning, Urban Geography, Physical Geography and Environmental Geography.

Springer Geography—now indexed in Scopus

Katarzyna Podhorodecka · Tomasz Wites
Editors

Global Challenges

Social, Economic, Environmental, Political and Ethical

Editors
Katarzyna Podhorodecka
Faculty of Geography and Regional Studies
University of Warsaw
Warsaw, Poland

Tomasz Wites
Faculty of Geography and Regional Studies
University of Warsaw
Warsaw, Poland

ISSN 2194-315X ISSN 2194-3168 (electronic)
Springer Geography
ISBN 978-3-031-60237-5 ISBN 978-3-031-60238-2 (eBook)
https://doi.org/10.1007/978-3-031-60238-2

This Springer imprint is published by the registered company Springer Nature Switzerland AG
The registered company address is: Gewerbestrasse 11, 6330 Cham, Switzerland

Introduction

We live in a constantly changing reality, in which global issues are redefined and require reliable interpretation based on the results of scientific research. The lack of reference to the science poses the risk of falsifying social perception in the multitude of feeding frenzy. Many global issues are phenomena and processes of a global nature and affecting most societies. They constitute existential challenges for the international community. The interest in global problems appeared at the turn of the 1960s and 1970s—the report of the 'Club of Rome' and U. Thant Report (op. cit.).

The geographical approach focuses mainly on the analysis of spatial and resource diversification. In terms of geography, one often relies on spatial analysis (i.e. the study of all spatial aspects, such as natural conditions, geographic location and socio-cultural conditions). Scientists use this approach by examining the principles of distribution, relationships and interactions that arise within it. An economic approach is to show the distribution of economic resources, production, distribution and consumption of goods and services. On the other hand, the sociological approach focuses on examining the systematic functioning and changes of society. The sociological approach examines social norms, processes and structures between people.

The aim of this book is to present selected contemporary issues connected with human development in the world. The selection of various issues, important in the modern world, is distinguished by great versatility and a wide thematic context. These problems have been divided into several groups which will be analysed in a geographic perspective. In the first group are problems connected with the environment; in the second are problems connected with the social public sphere; in the third, economic issues, in the fourth selected political issues and finally in the fifth, ethics issues of the contemporary world. The most important environmental problems include biodiversity loss, habitat destruction, the international trade of endangered species and natural disadvantages connected with tourism, especially in less developed countries. However, the solutions for the environmental issues can be a sustainable development. This way of development it presented in tourism example taken from the Alp regions experience.

In the previous decade, the consequences of the COVID-19 pandemic, which had begun in 2019, became a huge challenge starting from 2020 and continues to be a challenge in 2021. This is why this book has been updated with the newest indicators and data analyses. Moreover, the start of war in Ukraine February 2022 brought a huge social and economic disturbance to many European countries. Among the social problems are international migration and changes connected with pandemic situation in recent years. The book will also present selected political problems from the perspective of Central and Eastern Europe, concerning decommunizing public space. They have been selected as examples to illustrate the political issues of iconoclasm in respect of Warsaw monuments from the 1945–1989 period and post-Soviet monuments in Ukrainian cities. Economic problems included the influence of the global economic crisis and COVID-19 pandemic on the pace of global economic growth and the development of the tourism sector, the regional diversification of the global economic crisis and COVID-19 pandemic. The analysis of the global problem of tax havens and the problem of selling the citizenship by selected countries was shown. The last two chapters of the bool are showing the ethics problems connected with the natural hazards and disasters within the scope of geo-ethical problems with the concept of disaster risk reduction.

This book allows not only to get acquainted with description of the selected global issues in the contemporary world from the geographic perspective, but also allows to understand the complexity of the processes and dependencies taking place. Thanks to this publication, the public will be solidified in its conviction that geography is one of the meaningful parts of science which has a significant contribution in explication of real-life problems. Additionally, the problems addressed in this book have been illustrated with graphs from the areas under discussion, as well as with figures presenting statistical data relating to the environmental, economic, social and ethical problems.

Interpretation of the environmental, social and economic changes requires a precise presentation of the mechanisms shaping the contemporary view and forecast changes in the future. We hope that this publication will give you such an opportunity and will meet with interest of a wide range of readers—students of environmental, social and economic courses, journalists and anyone interested in understanding the complexity and intricacy of the contemporary world. This book is especially dedicated to the Erasmus students, who are joining the geography path.

The Global Shapers Organization, which was created by the head of the World Economic Forum, conducted a survey among millennials from around the world. A total of 26,000 young people from 181 countries around the world to identify the global problems they face nowadays. The study showed that the most important global problems among this generation are climate change, poverty and armed conflicts. Below is the ranking of upcoming global problems among millennials with the percentage of responses: (1) Climate change (45%), (2) Military conflicts (38%), (3) Religious conflicts (34%), (4) Poverty (31%), (5) Corruption, lack of transparency of governments (22%), (6) Lack of security and prosperity (18%), (7) Lack of education (16%), (8) Political instability (15.5%), (9) Lack of food and water (15,1%), (10) Unemployment and economic problems (14%) (*10 największych problemów współczesnego świata zdaniem millenialsów, [10 biggest problems of the modern world according to millennials, 2018*]). So, the list can be divided into the problems connected with the environment (climate change, lack of food and water), social problems (unemployment, religious conflicts, lack of education), economic problems (poverty, corruption, lack of security and prosperity) and political problems (military conflicts, political instability, lack of transparency of governments). Of course, in this book it is impossible to describe all of the global problems that is why the selection was taken from geographical perspective connected with the social-economic and environmental geography.

The book was prepared by authors working and associated with University of Warsaw, Faculty of Geography and Regional Studies, Chair of Regional and Political Geography.

Anna Dudek[1]—doctor, regional and environmental geographer; she completed doctoral studies at the Faculty of Geography and Regional Studies at the University of Warsaw and master studies at the Inter-Faculty Studies in Environmental Protection; her research interests include: environment and socio-economic development, determinants of nature conservation, tourism in protected areas, social and economic functions of national parks and reserves, development projects in deserted areas, environment and humanitarian aid; she conducted field research in many regions of the world, mainly in Africa, the Middle East and Latin America; she is the author of numerous publications on protected areas, nature tourism and development projects.

Filip Florczak—graduated student, Faculty of Geography and Reginal Studies, University of Warsaw.

Magda Głodek, Białek—graduated student, Faculty of Geography and Reginal Studies, University of Warsaw—specialization—geography of the world.

Ada Górna[2]—doctor and urban geographer at the Faculty of Geography and Regional Studies at the University of Warsaw, at the Chair of Urban Geography and Spatial Planning. Her doctoral thesis (2021) discussed the role of urban agriculture in the spatial and functional structure of the cities of the Global South. She is particularly interested in the issues such as food systems, urban sustainability, urban political landscape as well as spatial development, and conducted fieldwork concerning these topics in, i.e., Singapore, Kigali, Havana and Bissau. She is the author and co-author of many scientific publications, including those in prestigious journals such as: *Land*, *Journal of Baltic Studies* and *Miscellanea Geographica*.

Krzysztof Górny[3]—doctor and political and historical geographer at the Faculty of Geography and Regional Studies at the University of Warsaw. His doctoral thesis (2020) involved the colonial heritage in contemporary urban space of West African cities. He is currently working on a project focusing on Portuguese colonial heritage in the urban space of West African countries, and he conducted fieldwork on the topic in Cape Verde, Guinea Bissau, Ghana and São Tomé and Príncipe. He is also

[1] https://orcid.org/0000-0001-8336-9468, Faculty of Geography and Regional Studies, University of Warsaw.

[2] https://orcid.org/0000-0003-1591-7260, Faculty of Geography and Regional Studies, University of Warsaw.

[3] https://orcid.org/0000-0002-8088-3389, Faculty of Geography and Regional Studies, University of Warsaw.

interested in the political landscape of post-Communist cities of CEECs. He is the author and co-author of over 20 publications, including those in prestigious journals such as: *Planning Perspectives*, *Journal of Urban History* and *Journal of Baltic Studies.*

Barbara Dominika Jaczewska[4]—doctor in Earth Sciences in the field of socio-economic geography at the Faculty of Geography and Regional Studies. She is working as an assistant professor at the University of Warsaw since 2012. She conducts research related to social geography and spatial management; in particular, she is interested in issues related to spatial mobility, migration and integration, social and spatial segregation, and multilocality and housing. Most of her research was focused on countries in the European Union (especially Germany and Poland). Currently, she is conducting research on residential multilocality in Poland and its importance for a sustainable spatial management. She is an author or co-author of more than 30 publications, including scientific and academic books, articles and reviews.

Katarzyna Podhorodecka[5]—Associate Professor, Ph.D. with habilitation at the Faculty of Geography and Regional Studies at the University of Warsaw and studies at the Warsaw School of Economics. She obtained the academic title of habilitated doctor in the discipline of socio-economic geography and spatial management and published her habilitation thesis entitled 'Tourism in island areas in the face of the global economic crisis'. She worked at the Ministry of Development, and currently at the Ministry of Sport and Tourism, dealing with research in tourism, national tourism development, as well as competitiveness and innovation in tourism. She is the author and co-author of over 50 publications, in including in prestigious journals such as: *Islands Studies Journal* and *Current Issues in Tourism.*

Dorota Rucińska[6]—doctor in Earth Sciences in the field of geography (2009), assistant professor at the Faculty of Geography and Regional Studies at the University of Warsaw since 2010. She led to publication of a book *Extreme Natural Phenomena and Social Awareness* (2012). Her particular scientific interest is in nature–society relationships as social vulnerability to natural hazards (as multihazards) that are important for the disaster risk reduction implementation. She is focusing at social vulnerability and risk estimating at the local level. She has published in international scientific journals (e.g. *the Risk Analysis*, and the *International Journal of Disaster Risk Reduction*). In years 2014–2019, she initiated and built an international scientific network for conferences 'Disaster Risk Reduction'. She is interested in North America region especially the USA and developed the social vulnerability estimation at the local level and also interested nature–society relationships because of shale

[4] https://orcid.org/0000-0002-3911-7885, Faculty of Geography and Regional Studies, University of Warsaw.

[5] https://orcid.org/0000-0003-3075-7453, Faculty of Geography and Regional Studies, University of Warsaw.

[6] https://orcid.org/0000-0002-6493-3833, Faculty of Geography and Regional Studies University of Warsaw.

gas exploitation and ethics problems. She is interested in wide spectrum of the nature protection in the USA and also culture diversity in this country.

Tomasz Wites[7]—Associate Professor, Ph.D. with habilitation. Research issues investigated as part of the geography of population and social geography. The area of research interest is the regional scale above all (Russia in particular, other former USSR states to lesser extent, some other countries—New Zealand). To a limited extent the area of reference will be the global scale (globalization processes) and local scale (analysis of social pathology). He indicates several groups of research interest: those connected to the issue of population geography, depopulation in particular, related to social geography and with ties to tourism geography.

Katarzyna Podhorodecka
Tomasz Wites

[7] https://orcid.org/0000-0003-3161-8551, Faculty of Geography and Regional Studies, University of Warsaw.

Contents

Abbreviations

COVID-19	Coronavirus disease
CPI	Corruption Perceptions Index
DRR	Disaster risk reduction
GDP	Gross domestic product
IAS	Invasive alien species
IBC	International Building Code
IMV	International medical volunteering
IPBES	Intergovernmental Science-Policy Platform on Biodiversity and Ecosystem Services
NPO's	Non-Profit Organizations
OECD	Organization for Economic Cooperation and Development
RRF	Recovery and Resilience Facility
SIDS	Small islands developing states
TCM	Traditional Chinese medicine
TDRR	Triangle of Disaster Risks Reduction
TVE	Vertical evacuation
UNEP	United Nations Environmental Programme
WCDR	World Conference on Disaster Reduction

List of Figures

List of Tables

Chapter 1
Selected Environmental Issues in the Contemporary World

Anna Dudek, Katarzyna Podhorodecka, and Filip Florczak

1.1 Introduction

In this chapter, habitat and biodiversity loss; examples of island areas that have suffered from loss of habitat and species; and natural, cultural and economic disadvantages connected with tourism in less developed countries is covered. However, the development of human activity in the contemporary world has brought the term of sustainable development. There are presented the selected destinations that before COVID-19 were overcrowded such as: Caribbean Region (especially Cayman Islands), Maldives or Thailand Island Koh Phi Phi. In this chapter it is presented additionally on the example of tourism sustainable development of the Alps Region (France, Switzerland and Slovenia) as a remedium for uncontrolled the development of tourism infrastructure and tourism movement.

1.2 Habitat and Biodiversity Loss

This chapter addresses the natural problems of the contemporary world in developing countries that are related to the loss of a large areas of habitat, and examples of island areas where greed of human activity has contributed to the extinction of individual animal species. Isolated, insular areas were selected for the case studies: the Galapagos Islands, Mauritius and Madagascar. Additionally, the isolation index was analyzed, and its potential for application is shown in determining the level of isolation of islands in relation to the threat of colonization by individual invasive species of plants and animals.

A. Dudek (✉) · K. Podhorodecka · F. Florczak
Faculty of Geography and Regional Studies, University of Warsaw, ul. Krakowskie Przedmieście 30, 00-927 Warszawa, Poland
e-mail: adudek@uw.edu.pl

K. Podhorodecka and T. Wites (eds.), *Global Challenges*, Springer Geography,
https://doi.org/10.1007/978-3-031-60238-2_1

Unfortunately, climate change is not the only threat to the natural world, or in particular, to the number of individual species living in the world (www.naukaw polsce.pap.pl). According to Morelli et al. (2020), the greatest threat to the animal world, apart from climate change, is the degradation and fragmentation of habitats. Usually, this is led by a wasteful economy, hunting and the occurrence of invasive species and high levels of environmental pollution. The authors analyzed the example of the island of Madagascar, which has experienced extensive deforestation and over-farming. This has had a negative impact on the natural environment and is still a huge threat to the species of lemurs that live there.

Loss of biodiversity and habitat loss are the greatest threats to the proper functioning of life across the world (www.wfos.gdansk.pl). Each species has a specific function in the ecosystem and has its place. If the specie's numbers are reduced or if it dies out, the ecosystem becomes less stable. This means that living organisms influence each other and: this results in a network of food deficiencies because of matter circulation and energy flow take place in nature (www.wfos.gdansk.pl). The loss of biodiversity will have a catastrophic effect on humans worldwide (www.zie lonewiadomosci.pl; www.ipbes.net).

According to the *Report of the Intergovernmental Platform for Biological Diversity and Ecosystem Functions* (IPBES), in which the impact of economic development on nature and ecosystems was analyzed, the current rate and scale of species extinction have never been observed before in the history of mankind (www.naukaw polsce.pap.pl). The last five decades have been critical to the rate of extinction, as the world's population has doubled in that time. At the moment, one million species of all classified species of plants and animals (excluding microorganisms) is at risk of complete extinction (www.newsweek.pl).

It should also be remembered that climate change affects, and will affect, the range and distribution of species as well as their reproductive cycles or vegetation periods (www.klimada.mos.gov.pl). This means that some species may gain, but many may lose. Losses will definitely be recorded in the numbers of species, especially among those that are endangered. Table 1.1 presents the classification of species according to adopted threat categories, and including sample names of species.

Habitat loss is primarily caused by humans. This is due to land clearing for farming, mining, grazing, drilling and urbanization indeed. These results that about 80% of global species need to be in the forest as a main habitat. Each year it is estimated that 15 billion trees are cut down (www.blog.nationalgeographic.org). Researchers analyze, most often, the declining areas of various types of forests, but it must be remembered that other habitats are also declining: wetlands, heaths, peatlands, grasslands, etc. A habitat for living species is a set of mainly climatic and soil factors, which are independent of biocoenosis and that prevail in a specific place, and affect the development of individual organisms. Natural conditions affect their populations and the entire biocoenosis system. A given habitat determines the conditions for the existence of individual types of plant communities and related animal groups. The following types of habitats have been distinguished: natural habitats; synanthropic habitats that have arisen as a result of human activity, including ruderal

Table 1.1 Classification of species according to adopted threat categories

Abbreviation	Explanation	Information and examples
EX	Extinct	Extinct species—dodo bird
EW	Extinct in the wild	Species extinct in the wild—this means that they live only by being bred by humans, i.e., in zoos: Przewalski's horse, sable oryx
CR	Critically Endangered	Extremely endangered species: Algae lemur, eastern gorilla
EN	Endangered	Endangered species: lemur katta, Siberian tiger
VU	Vulnerable	High-risk species, threatened with extinction due to progressive population decline: crayfish, African manatee
LR	Lower risk	Lower risk species
NC	Near threatened	Lower risk species, but fairly close to being threatened—i.e., European otter
LC	Least concern	Species requiring the least care, but showing no clear population regression,—i.e., mud bithynia
DD	Data deficient	There are insufficient data
NE	Not evaluated	No assessment was done

Source elaboration based on the *Red List, ICUN Red List Categories and Criteria*, Version 3.1, 2000, IUCN Species Survival Commission, CouncilGland, Switzerland, www.iop.krakow.pl

ones; segetal habitats; and semi-natural habitats. The destruction of natural habitats by a destructive human economy has huge consequences for the abundance of individual plant and animal species.

The fact that habitats are shrinking is a serious problem for biodiversity. Nevertheless, another type of threat is the destruction of habitats in the form of their fragmentation. Initially, the degraded habitats are relatively continuous and the populations living there may still be in contact with each other without hindrance, but over time the loss of habitat reaches such levels that isolated patches of habitats are formed along with isolated populations within them that lose contact with each other. Such situations bring many additional threats, both for the entire species complex and for individual populations (Pullin 2004).

Mechanisms contained in the theory of island biogeography are used to explain the effects of habitat fragmentation. It is assumed that a fragment of the habitat is a kind of island. As the area of such a fragment decreases and its isolation increases, there is a gradual decline (extinction) in species, and the chances of other species colonizing the fragment decrease.

Another problem with habitat fragmentation is the increase in the amount of edge area in relation to the total area of the habitat. According to researchers, 70% of forests in the world are located within 1 km of their edges, and 20% of forests are only 100 m from their edges (Haddad et al. 2015). This creates a whole range of threats to the populations that live there. Edge habitats are very different from internal habitats and often become unsuitable for some species. At the edges of habitats, different microclimatic conditions prevail, and they are exposed to a greater frequency of environmental disasters and more often penetrated by predatory or competitive species. With greater edge are in relation to their total area, the probability of passive migration from the habitat is also greater. An increase in edge area translates into lower biodiversity (Pullin 2004).

Habitat fragmentation also means the fragmentation of a specific population of a specific species. The remaining, less numerous populations, are more likely to become extinct, for several reasons. First, a population may become extinct as a result of the random variation in birth-to-death ratios. In large populations, fluctuations in the number of individuals will be small relative to the size of the population, but if the population is small this situation may create a situation in which there is a risk of extinction. Second, in small populations the chances of finding a reproductive partner are lower. Moreover, group protection against predators is reduced, the efficiency of group feeding is reduced, and in extreme cases the social hierarchy and social bonds may break down. Another threat to small populations is the random loss of genetic variation, and, related to this phenomenon, more frequent appearances of individuals with genetic abnormalities. Small populations are also more vulnerable to environmental disasters. Any unfavorable weather change, flood, fire, or volcanic eruption may affect the entire population and lead to its extinction (Pullin 2004).

The above-mentioned threats make it very important for connectivity to be maintained between the core zones of habitats remaining in the environment. A network of such connections in the form of ecological corridors, stepping stone habitats, or ecoducts (wildlife crossings over or under roads, highways and railroads) offers the possibility of wild animal migration and gene exchange.

Another important environmental problem contributing to the reduction of biodiversity is the spread of invasive alien species (IAS) (i.e., non-native species that are alien to a given ecosystem). These pose a threat to local species of fauna and flora, by occupying their ecological niches, competing with them, feeding on them, changing their environment, or interbreeding with them (hybridization). It is estimated that since the seventeenth century, invasive species have contributed to 40% animal extinction cases (www.cbd.int/idb/2009/about/what). Invasive species have features that allow them to win the fight against native species, including: large coverage areas, rapid growth and reproduction, tolerance of various environmental conditions, and the easy colonization of new areas, often with human participation. The three most common routes through which such species spread are international trade, transport and the deliberate introduction of alien species for cultivation, breeding or pest control in agriculture (*The Nature Conservancy*). Invasive alien species pose a particular threat to island species as they are usually not suited to competing with newcomers.

According to Bhuiyana et al. (2018) per capita food supply is influencing with the N_2O emissions. Moreover, the GDP indicator and trade liberalization policies are also impacting on these environmental emissions, along with NOx emissions, in the region.

The following part of the chapter describes island areas that have been characterized by loss of habitat and endemic species.

1.3 Examples of Island Areas that Have Suffered from Loss of Habitats and Species

A special isolation indicator has been developed by the United Nations Environmental Program (UNEP) to measure the degree of insulation. It measures the degree to which an island is isolated from potential sources of colonization by living organisms. This indicator is 'the sum of the square roots of the distance of the island to the nearest comparable or larger island, to the closest group of islands or archipelago, and to the nearest continent.' If there is no continent nearby, the greatest distance is repeated except for small satellite islands, which are close to large land (United Nations Environment Programme, isolation indicator—www.islands.unep.ch). The UNEP isolation index was designed to determine the degree to which an island territory is isolated, in relation to its potential for colonization by individual plant and animal species.

Table 1.2 presents the UNEP isolation index for selected island areas.

The world's island territories are extremely isolated areas; this means that a significant number of endemic species are found on them. Endemic species are those living organisms that only exist in a given area and cannot be found elsewhere in the world.

1.3.1 The Galapagos Islands

The Galapagos Islands archipelago is located in the Pacific Ocean on the equator, about 1000 km west of the coast of South America, and has an area of 8000 square kilometers. It is an unusual area characterized by a large number of endemic species. The archipelago belongs to Ecuador. The islands are famous due to Charles Darwin's nineteenth-century expedition to visit the islands. He arrived in 1835 and did research on four islands: San Cristóbal, Floreana, Isabela and Santiago (Porter and Graham 2015).

In 1959, almost the entire land area of the Galapagos Islands was included in a national park, and in 1998, the waters surrounding the archipelago were also protected as part of the Galapagos Marine Reserve. In 1978, the archipelago was inscribed on the UNESCO World Heritage List, and in 2001, the entry was extended to include the Galápagos Marine Reserve. More than 80% of all land birds, 97% of reptiles and land mammals, more than 20% of marine species, and 30% of all plants found on the islands are endemic species. Tourism is the largest source of income for these islands but at the same time represents its greatest threat. The Galapagos Islands are visited by approximately 250 thousand tourists each year (Fig. 1.1).

From the beginning of its discovery in the sixteenth century, the archipelago has been of interest to various groups of people: sailors, whalers, pirates, settlers, and, today, also scientists and tourists. These first groups explored the islands and exploited the natural resources, leading to the decimation of the populations of native fauna species. This was the case, for example, with the Galapagos giant tortoises,

Table 1.2 Isolation indicator for chosen island areas, according to UNEP

Isolation indicator	Island—Country	Island distance	Group distance	Continent	Distance (km)
61	Pinzon (Ecuador—Galapagos Islands)	10	850	South America	850
62	Efate (Vanuatu)	75	340	Australia	1200
63	Epi (Vanuatu)	22	550	Australia	1200
64	Ranongga (Solomon Islands)	10	500	Australia	1500
65	Iceland (Iceland)	300	300	Europe	950
66	San Cristobal (Ecuador—Galapagos Islands)	65	850	South America	850
67	Uoleva (Tonga)	3	50	Australia	3400
68	Ontong Java (Solomon Islands)	60	250	Australia	2000
69	Mussau (Papua New Guinea—Bismarck Archipelago)	95	550	Australia	1300
70	Hunga Ha'apai (Tonga)	4	100	Australia	3400
71	Cook (United Kingdom—Falkland, S. Sandwich)	1	800	Antarctic	1750
72	Salomon (United Kingdom—Chagos Archipelago BIOT)	30	600	Asia	1750
73	Reunion (France—Reunion)	150	700	Africa	1200
74	Faial (Portugal—Azores)	6	1100	Europe	1500
75	Alofi (France—Wallis and Futuna)	5	260	Australia	3200
76	Frigate (Seychelles)	25	1100	Africa	1400
77	Helen (Palau)	75	300	Asia	2600
78	Saipan (United States—Northern Mariana Islands)	10	500	Asia	2800
79	Aguijan (United States—Northern Mariana Islands)	5	500	Asia	3000
80	Nanumanga (Tuvalu)	50	400	Australia	2800
81	Hunga Ha'apai (Tonga)	55	300	Australia	3200
82	Corvo Island (Portugal—Azores)	19	1300	Europe	1700

(continued)

Table 1.2 (continued)

Isolation indicator	Island—Country	Island distance	Group distance	Continent	Distance (km)
83	Beqa (Fiji)	10	800	Australia	2700
84	Ata (Tonga)	150	300	Australia	3000
85	Auckland (New Zealand—Auckland Islands)	400	460	Australia	1900
86	Arorae (Kiribati)	90	300	Australia	3500
87	Kanacea (Fiji)	15	800	Australia	3000
88	Aranuka (Kiribati)	5	700	Australia	3500
89	Bounty Islands (New Zealand—Bounty Islands)	200	600	Australia	2500
90	Futuna (France—Wallis and Futuna)	300	260	Australia	3200

Source United Nations Environment Programme (www.islands.unep.ch/Tisolat.htm)

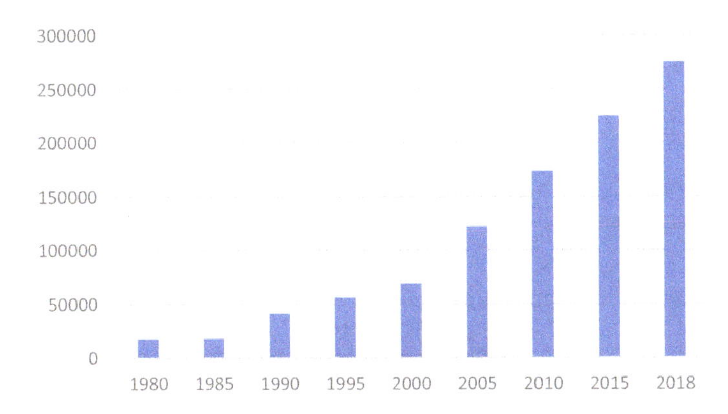

Fig. 1.1 Number of tourists visiting the Galapagos National Park in 1980–2018. *Source* www.galapagos.gob.ec (18.06.2021)

which were harvested as 'live refrigerators' and loaded onto ships as a source of food for the crew. Their fat was also used as oil for lamps in the cities of mainland Ecuador. The species survived thanks to conservation efforts (i.e., breeding centers) and the establishment of the national park.

Other groups of settlers brought invasive species to the Galapagos Islands. Some came here accidentally (i.e., rats, insects), others deliberately (i.e., dogs, cats, goats). To this day, the presence of invasive alien species is considered one of the greatest threats to the flora and fauna of the archipelago. The national park staff conduct many actions and allocate huge funds to the eradication of unwanted migrants ants, snails, flies, rats, goats, stray dogs and cats, etc. These species pose a threat to native species because, depending on the genus, they can compete with them for the same ecological niche, feed on the eggs or hatchlings of endemic species of birds and reptiles, and can also parasitize them. The scale of their impact is, therefore, very large. Island native species that evolve in relatively isolated conditions are usually unfit to compete with strong alien species and are therefore forced out of their natural habitats.

Human settlement has also been a threat to the nature of the Galapagos Islands. Especially in the 1990s, population migration from the continental part of Ecuador reached about 8% per year. The immigrants were looking for jobs in the fishing industry or tourism. Uncontrolled migration, settlements and the expansion of urban areas were the reasons why the Galapagos National Park was inscribed on the List of World Heritage in Danger. For the moment, the park is no longer listed on the List because the authorities have managed to limit migration from the continent. Nowadays, tourism is also a threat. Due to the islands' natural uniqueness, geological history, and faunal endemism, the islands attract more and more tourists, both Ecuadorians and foreign. This results in the development of tourist infrastructure: hotels, diving centers, tourist agencies, transport infrastructure, as well as an increasing number of ships cruising the Galapagos Islands. The tourism industry also produces waste and pollution, and tourists either scare wild animals or make them accustomed to seeing people. Tourist arrivals are also a potential factor in contributing to the introduction of new, alien species of animals and plants from the South American continent.

1.3.2 Mauritius

Mauritius is an island group with an area of 2045 km^2 and a population of 1.26 million (2022) (www.data.worldbank.org). These islands were uninhabited, when the first colonizers came to the area. The dodo bird lived on Mauritius until the seventeenth century, but became extinct somewhere between 1662 to 1693; the exact date of extinction unknown. Predators brought to the island by colonizers were responsible for the disappearance of the dodo bird from the central part of the island of Mauritius. The bird had no enemies before colonization and, therefore, was not particularly afraid of humans, or the dogs and cats brought to the island. The Dutch originally colonized Mauritius between 1598 and 1710. The island was characterized by very different colonial dependencies (French, British and Dutch). At present, Mauritius is inhabited by an ethnic and religious mix, and the dodo bird symbol is used in tourism marketing because it is recognizable all over the world as a species killed by human greed. In terms of the tourism economy, the area began to develop in the 1970s, and the economic importance of tourism on Mauritius has been growing steadily. In the 1970s, there were 25 hotels on Mauritius, and in the 10 years to 1982 the number doubled to 51, doubling again to 103 hotels in 2004, which means that from 1972 to 2004 the number of hotels in Mauritius increased more than four times.

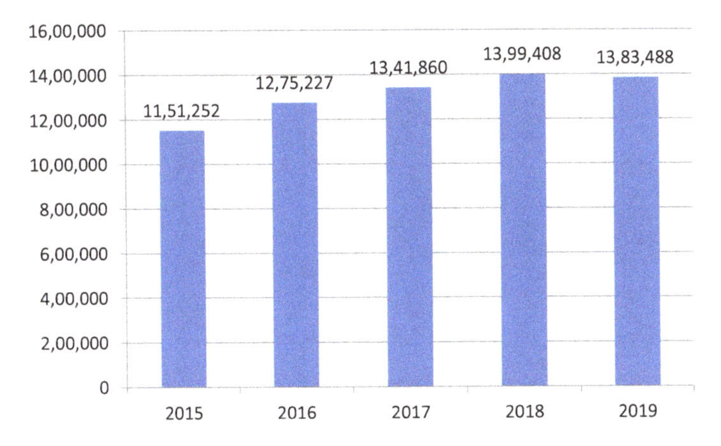

Fig. 1.2 Number of arrivals of foreign tourists to Mauritius in 2015–2019. *Source Yearbook of Tourism Statistics Data 2015–2019*, 2021, UNWTO, Madrid

Figure 1.2 shows the number of foreign tourist arrivals to Mauritius during 2015–2019. Before the COVID-19 pandemic, there was a significant, 20% increase in foreign tourist arrivals to Mauritius over the 2015–2019 period.

1.3.3 Madagascar

Madagascar is an island with an area of 587 thousand km^2 and is located in the Indian Ocean east of the African coast; it has a population of 25 million people. Here, there are species of lemurs that are endangered. Madagascar has significantly increased the pace of tourism development from 244 thousand foreign tourist arrivals in 2015 to 383 thousand in 2019. Before the COVID-19 pandemic in, the arrivals of foreign tourists between 2015 and 2019 increased by 57%.

Madagascar attracts tourists due to the unique nature of its natural environment, in particular the high number of endemic species. It has become a destination for nature tourism, but unfortunately there is no adequate tourism development plan, which means that tourist movement fluctuates, especially during periods of political instability (Jędrusik 2018). Figure 1.3 shows the number of foreign tourist arrivals to Madagascar during 2015–2019.

1.4 International Trade in Endangered Species

The contemporary increase in interest in buying exotic animals or parts of their bodies is due to the increased purchasing power of societies that have not been able to buy luxury goods until now. First of all, Chinese people are important buyers of such

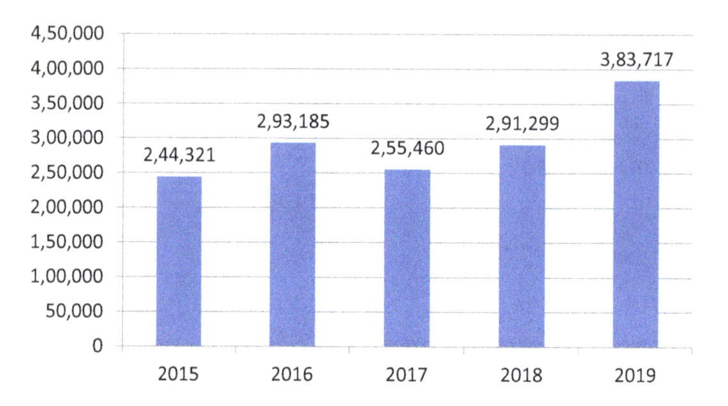

Fig. 1.3 Number of arrivals of foreign tourists to Madagascar during 2015–2019. *Source Yearbook of Tourism Statistics Data 2015–2019*, 2021, UNWTO, Madrid

goods. Many products obtained from wild animals are used in traditional Chinese medicine as well as in Chinese cuisine. These products are considered to be of medical importance and to have a beneficial effect on the body. This also translates into a high demand in countries with a large Chinese diaspora, for example, Singapore, Malaysia, Indonesia, etc. The increase in popularity and value of products obtained from wild animals reduces the medical importance of such products and the belief in their miraculous properties (typical for the older generations), and the real driving force is their investment value, fashion for owning or using certain things, and their importance as a determinant of social status. The popularity of such transactions is favored by universal access to the Internet, through which many such sales and purchase contracts are concluded—not always legally. It is also facilitated by the development of transport (including air and sea transport) and the possibility of sending goods over long distances. The development of sea transport by container ships makes it easier to send prohibited goods, because they can be hidden in a container among other products.

According to reports on wildlife trade, pangolins are among the most commonly smuggled animals today, and are valued in Asian markets for their meat and scales, which allegedly have medicinal properties. Although trade in all eight species of pangolins, four Asian and four African, is banned according to CITES (the Convention on International Trade in Endangered Species of Wild Fauna and Flora), there are many cases of illegal harvesting and consumption of their scales and meat. Other more famous cases include the bones and other body parts of tigers. In China, these are used to prepare wine with healing properties. Tigers are found in various parts of South and East Asia (around 4000 individuals), but the most are in captivity in special breeding farms in China (around 5000), where they are often starved or illegally killed for their skeletons, meat, and precious leather. Other ingredients used in Chinese cuisine include sea cucumber, which are is such great demand that they are grown in special aquaculture facilities in various countries, and then exported to China. Some species of sea cucumbers are even caught within the boundaries of marine reserves, such as the Galapagos Marine Reserve. Another delicacy in Chinese cuisine is swallow's nests made of the saliva of certain species of swiftlets (*Aerodramus fuciphagus* and *Aerodramus maximus*), which nest in dark and humid caves in Southeast Asia. Nutritional soups are prepared from these nests, which are popular not only in China, but wherever a large Chinese diaspora lives. In some caves, collecting nests takes place after the chicks have hatched, and then such practices are considered sustainable. This is the case at the protected Gomantong Caves in Sabah, Borneo. Twice a year, licensed harvesters are permitted to collect a certain number of nests there. A different situation concerns the practices at similar caves in the south of Thailand. These caves are on lease to large and influential companies that have licenses from the authorities to collect swallow nests there. Access to these caves is very difficult, but many environmental reports indicate that the nests are gathered before the chicks are hatched. The demand for swallow nests is so great that in some places (i.e., in Kumai in the Indonesian part of Borneo, and in Sekinchan in Malaysia), the Chinese are investing in swiftlet farming, for example, the construction of special structures that serve as artificial places for the nests. These buildings are like houses without windows; inside, the conditions that usually prevail in the caves are recreated. This practice is considered to be beneficial from an environmental point of view as it allows wild bird populations to reproduce peacefully in the wild in their caves. Moreover, the 'production' of nests in such places takes place in more hygienic conditions. It is a very profitable business, and it is not surprising that more and more countries are involved in such production, such as Cambodia and Vietnam.

The exotic pet trade is yet another market. Its concerns are mainly exotic species of birds (parrots) and reptiles (turtles, lizards, snakes), as well as mammals (monkeys, rodents) and amphibians. The animals are harvested for breeding activities, sold locally, or smuggled abroad. Often, they are deliberately mislabelled as captive bred and are then legally exported. Statistically, on average 72% of the animals do not survive their transport to the destination. This is due to dehydration, stress, infection, hunger, parasite infestation, cage overcrowding, trauma and poor hygiene (Ashley et al. 2014). The animals are captured from the tropical regions of Africa, Asia and Latin America, all identified as exotic pet trade hotspots. Hunting for wild animals not only contributes to a reduction in biodiversity, but also poses a sanitary and epidemic threat, as evidenced by the COVID-19 pandemic. Most infectious zoonoses, such as SARS, Ebola, MERS and COVID-19, probably come from wildlife.

A particular threat is posed by the trade in endangered species that are protected by CITES, but it must be remembered that the legal trade in species with more numerous populations also continues. It, too, poses a threat to public health and is a source of cruelty and animal suffering.

1.4.1 The Illicit Trade in Rhinoceros' Horn

It has long been believed that rhinoceros body parts have miraculous properties. For centuries in South Asia there were beliefs that everything that came from the rhinoceros, especially its horn and blood, had miraculous and healing properties. In Nepal, rhinoceros' blood and meat were sacrificed during religious ceremonies; in India, it was customary to drink the blood of a rhinoceros during religious celebrations, or at least drink water and milk from cups carved from the horn. Also in East Asia, mainly in China, there was, and still is, the belief that the rhinoceros' horn is a very effective medicine for fever, typhoid, headache, nausea and a number of other ailments. It was also used as a snake venom neutralizer. However, there are no scientific studies confirming the healing properties of rhinoceros' horn. But such a broad, alleged spectrum of activity has meant that the demand for this product has always been very high, and, therefore, the problem of the illegal trade in rhinoceros' horn has been discussed for a long time. It is a problem well known to the general public. The intensity of this phenomenon over the past 40 years, however, has varied. In the 1970s and 80 s, the hunting of rhinoceros, mainly in Africa, was very intense. The animals were killed in large numbers then, and their horns found their way to markets in South Korea, Taiwan, China and also Yemen (in the last country, rhinoceros' horns were made into the handles of traditional, male-worn daggers called *khanjar* or *jambiya*). The efforts of international conservation organizations led to a global ban on the trade in rhinoceros' horn in 1977. The removal of the rhinoceros' horn from the official list of ingredients used in traditional Chinese medicine (TCM) in 1993 also helped to control the intensive poaching of these animals. The animals were placed under strict protection and their populations in the protected areas were slowly recovering, even though poaching continued to occur, but was limited (Milliken 2014). The international community seemed to have coped with this problem. Everything changed in 2008. It began with a rumor that one former Vietnamese politician had been cured of cancer by ingesting powdered rhinoceros' horn. This story gained immense popularity, it crossed the borders of Vietnam, reaching China, where it only cemented the old beliefs about the healing properties of the horn. In Asia, this caused a sudden increase in demand for the alleged 'wonder drug'. This is especially true in Vietnam, where the death rate from cancer is currently increasing, and the queues for chemotherapy and radiotherapy treatment are very long. So, many people turn to other methods. In addition, the high price of the rhinoceros horn means that, for the Vietnamese, the opportunity to buy powdered horn is a symbol of wealth and status. Drinking water with powdered horn has become very fashionable at business meetings and family gatherings, especially in middle-class circles. So, since 2008, there has been a huge crisis in the area of horn smuggling. Every year, hundreds of animals are illegally killed, both in Africa and Asia. In 2015 alone, about 1,200 animals were killed by poachers in South Africa (www.rhinos.org/the-crisis). The gigantic demand has pushed horn prices to record levels: the most quoted rates are around 60,000 USD per kilogram, but can go as high as 100,000 USD per kilogram. Due to the efforts of environmental activists and national park rangers, the number of illegally hunted animals in South Africa has been decreasing for several years, but in 2019 it was still around 600 animals (www.rhinos.org; www.savetherhino.org).

Contrary to the above-mentioned, widespread beliefs, according to some researchers, the current demand for rhinoceros' horn in Asian countries, mainly in China, is due not so much to its medicinal properties, but to the fact that it is a good investment, and the industry that drives this the most is the art, antiques and jewelry market.

Of the five rhinoceros species surviving to this day, four are endangered, of which three are critically endangered. They are most numerous in Africa, which is why poachers are mainly active there. There are two species of rhinoceros that live in Africa. The black rhinoceros (*Diceros bicornis*), which is critically endangered. It occurs in protected areas in East and South Africa, with a population of between 5300 and 5600 individuals. The second African species, the white rhinoceros (*Ceratotherium simum*), is the least endangered; but, according to the *Red List*, it has near threatened status. Its numbers in the wild are between 17,200 and 18,900 animals. Most individuals are found in South Africa, primarily in the Kruger National Park. In addition, it lives in several other African countries, but in smaller populations. This species is the largest living rhinoceros and has the largest horns it is, therefore, most often the victim of poachers today. It is worth adding that although at the species level it is the least endangered rhinoceros, the northern subspecies *Ceratotherium simum cottoni* has already become extinct in the wild, with the last two individuals (two females) kept in captivity in the private Ol Pejeta reserve in Kenya. The three Asian rhinoceros' species are much less abundant. The most numerous is the greater one-horned rhinoceros (*Rhinoceros unicornis*), which has a population of about 3600 individuals, living in Nepal and northern India; it has a vulnerable status. The Sumatran rhinoceros (*Dicerorhinus sumatrensis*) lives in a few, fragmented populations in several national parks in Sumatra and most likely also in the Indonesian part of Borneo. It is critically endangered; only 80 individuals have survived to the present day. The last of the Asian species, the Javan rhinoceros (*Rhinoceros sondaicus*), is also a critically endangered animal. Nowadays, the last animals (74) are surviving in the Indonesian Ujung Kulon National Park in Java. Until recently, a second population of Javanese rhinoceros existed in the Cát Tiên National Park in southern Vietnam. Unfortunately, the huge demand for horn in this country meant that in 2010, a few years after the crisis started, poachers killed all the rhinoceros in this park.

The illegal trade in rhinoceros' horn is global. This problem is not unique to African or Asian countries. In 2011, very soon after the onset of the current crisis, cases of the theft of horn from natural history museums, zoos, private collections and auction houses occurred in many European countries. The people organizing the horn smuggling realized that it was much easier to organize such a theft in Europe than to hunt rhinoceros in Africa: many museums who had rhinoceros' horns were not guarded well enough, and, African wild populations became less numerous and better protected. In just two years, 2011–2012, Europol registered 67 such thefts and 15 unsuccessful attempts.

1.4.2 Ivory Smuggling

As in the case of rhinoceros, the 1970s and 1980s were also the most intense period for elephant hunting. At that time, Europe was the largest importer and processor of ivory. To protect elephant species from extermination, a global ban on the ivory trade was introduced in 1989. At the request of several African countries, the ban on the ivory trade was relaxed in 1997, legalizing the sale of ivory in some African countries (Botswana, Namibia, Zimbabwe, and, later, also South Africa). The precondition for loosening the law was that the ivory could only come from previously stockpiled supplies, not from the elephant population living in the wild. In Western societies, an awareness about the ivory trade has arisen, leading to a decline in demand for such products. Unfortunately, in recent years there has been an increase in the number of cases of poaching and ivory smuggling (Milliken 2014). Currently, the African elephant population is estimated to be 500,000 individuals (www.elephantconservation.org).

The reason for this contemporary increase in interest in ivory is the growth in the wealth of the societies of the Far East, where ivory products are very popular. Ivory smuggling routes start at the places where elephants are hunted and where their tusks are cut off. It is then transported overland to ports in West or East Africa, and there it is usually placed in containers that are shipped to Asia. The sea route is most often chosen because ivory is heavy and it is easier to hide it among the other goods in containers (Milliken 2014). The Asian transit port is most often Hong Kong, from where the tusks are transported to Chinese ivory processors. This raw ivory also goes to Thailand, the Philippines, Taiwan and Japan (Milliken 2014). It is worth noting that it is not only Asian countries that generate today's increased demand for ivory; it also occurs in African countries due to the presence of foreign investors (including Chinese) working in the mining or timber industries, or in large infrastructure projects, which translates into greater demand in Africa itself. Tusks are used to make sculptures, jewelry and religious items. Europe is not the main market for ivory today, but a relatively amount is sold in some countries, such as the UK and Germany. The USA remains the second largest market after China. The ivory that is sold there comes from legal sources (those imported before 1989) but also illegal—sold over the Internet.

1.5 Natural, Cultural and Economical Disadvantages Connected with Developing Tourism

In this chapter have been shown the natural, cultural and economical disadvantages connected with developing tourism in less developed countries especially the term of tourism neo-colonialism in less developed countries.

1.5.1 Tourism Neo-colonialism in Developing Countries

The development of tourism in developing countries is often called 'the tourism neo-colonialism' of the countries 'of the North' to the countries 'of the South' (Cywiński 2019). Tourism industry in less developed countries was to accelerate development, improve the economic situation and create new jobs. This is due to the fact that tourism seems to be a low-capital and highly labor-intensive sector of the economy. It was also hoped that tourism would improve the balance of payment and increase export volume.

However, famous tourist destinations are not getting the benefits they should. Southern countries use natural environment resources leading to degradation of the natural environment through the uncontrolled development of tourism. In addition, most of the profits from tourism in developing countries are transferred abroad by large hotel chains and tourism corporations (Barnwell 2003). Relying too much on tourism Mono-culture economy can lead to huge problems (Jasiński 2006). This may be clearly visible especially during the COVID-19 pandemic times.

According to Cywiński (2019) 'tourism neocolonialism is an asymmetrical relationship between the world of tourists and the world of local communities, which results also (which may take the form of subordination) and/or marginalization (which may take the form of exclusion) of the world of inhabitants by the world of tourists. In addition to the proposed definition, its individual elements were analyzed and the possible path of research on the phenomenon of tourist neocolonialism was explained, divided into its specific contexts corresponding to various geographical and social manifestations. We can show the tourism neo-colonialism is shown in four main aspects: economic, cultural, social and spatial. Tourism contributes to the increase in imports of developing countries, so that they can bring everything that tourists are used to (a large number of products are not available on local markets). The development of mass tourism causes negative social and cultural changes among the local population. Consequently, rich countries benefit more than poor countries from developing of tourism in their area.

Negative employment consequences of tourism development in Southern countries. The local population finds employment in low-wage and low-skilled jobs. Employed people cannot use medical care, they have unsatisfactory working and accommodation conditions. Negative features of tourism development in the 'countries of the South' related to employment. After the end of the season, local workers are usually dismissed without ensuring that they will find employment again next year. Employees do not receive training to improve their qualifications. Managers come from developed countries, which means that a large part of wages is paid outside borders of the country receiving tourists. The tourist expenses incurred in the destination country are largely transferred back to the countries of the North (i.e., through the purchase of imported products, the use of hotels belonging to international chains, etc.). For the majority of organized tourism offers 'all-inclusive package tours'. The number of transfers to rich countries is about 80% of revenues from tourism. A large part of tourists' expenses is spent on: expenses on airlines, travel agencies, foreign promotion, hotels of international chains (which are often based in the country, where tourists come from). For this reason, little money is left for local entrepreneurs and workers. This kind of tourism neo-colonialism was confirmed by Cywiński (2015), who analyzed effects of tourist neo-colonialism on local cultures and communities in Indonesia. The COVID-19 pandemic time is an exception from the profitable decade for tourism after the global economic crisis. During the COVID-19 pandemic people in tourism industry are losing their jobs very quickly, because of short time type of working agreement in tourism industry. The tourism economy in many places all over the World is closed because of restrictions and because of the scare of traveling.

1.5.2 Tourism Monocultures and the Influence of COVID-19 Pandemic on Tourism Sector

Being a tourism monoculture is not a good option. It can be observed a huge economic downturn during the global economic crisis from 2007 to 2010 and during COVID-19 pandemic 2020–2021. According to the UNWTO estimations the international tourist numbers are going to fall in 2020 from 60 to 80% (*Impact Assessment of the COVID-19,* UNWTO 2020). As a consequence of reduced tourist movement, the number of people traveling abroad will drop by 0.85–1.1 billion, which will bring a loss of export revenues at the level of USD 0.91–1.2 trillion. As a result of the crisis 100–120 million people are directly at risk of losing their jobs (www.unwto.org). Many countries all over the world are relying on tourism. This situation is observed in SIDS—small islands countries and dependencies in developing states. Tourism as the industry most affected by the COVID-19 pandemic. In the world tourism suffers its deepest crisis. In 2020 with a drop of 74% in international arrivals (UNWTO 2021). As a result of the introduced restrictions, there was a significant decrease in the number of people using the accommodation facilities all over the world and especially in developing countries.

1.5.3 Harms to the Environment by Tourism Industry

Tourism movement bring threat to the natural environment, especially in developing countries (Kruczek 2011). According to the definition of the Federation of National Parks and Nature Reserves of Europe 'sustainable tourism' is 'any form of tourism development, management and tourism activity that maintains the ecological, social and economic integrity of the areas, and preserves the natural and cultural resources of these areas unchanged' (Niezgoda Jantczak and Patelak 2020, after *Parks for life..., IUCN*, 1994; Zaręba 2000). In many regions of developing countries, the uncontrolled development causes huge problems to the natural environment, social and economic sphere.

Sustainable tourism should minimize the negative impact of tourism on the natural, economic and social spheres. Using this solution tourists should enjoy the tourist attractions limiting the degradation processes and preserving the attitudes for future generations (Niezgoda et al. 2020).

Many examples can be shown of huge exploitation of the natural environment by the tourism sector. It can result in water pollution, air pollution and the reduction of forest areas. Construction of tourist infrastructure near the seaside are destroying the environmentally valuable areas. The problem of tourism in mountain areas was shown by Čech and Javorska (2023).

1.5.4 Harms to the Culture, Economy and Spatial Development by Tourism Industry

Local culture is suffering because of tourism development. We can observe the collision of Western culture with the culture of the Islamic or Buddhist countries. Not everything is for sale. So often artificial folklore and culture is created for tourists. According to a study by Shakeela and Weaver (2018), in which the attitudes of residents toward the development of tourism in the Maldives are described, local residents saw tourism as the 'evil' from which their community should be 'isolated'. In the case of this island area, a specific isolation takes place because tourist islands are like separate areas, where locals do not locate their homes (Trykowski, Podhorodecka 2019).

1.5.5 Case Studies from Overcrowded Tourism Destinations

- **Caribbean Region and cruises ships industry**

The example of over tourism phenomenon and to high-level tourists' exploration was the Caribbean Region. It was especially due to cruise ship tourism industry just before COVID-19 pandemic times. The cruise ship industry was developing for the last 5 decades. It was estimated that in 70 s last century about 0.5 millions of people annually travel with cruise ships (www.businessinsider.com.pl). The growth rate of cruise tourists in the 80 s was 8% per year, twice the growth rate of tourists in the world as a whole (Dwyer 1998). In the 80 s and 90 s last century over 8.5 million passengers were transported annually on cruise ships (approximately 80% of cruise ships passengers were Americans) (*Tourism 2020 Vision. Global Forecast* 2001). In the late 1990s, 10 million cruise ships passengers were registered annually on ships operating in the Caribbean Region. As much as half of all cruise-ship passenger were in the Caribbean Region.

Until the COVID-19 pandemic cruise ship industry was a very dynamically developing field of tourism. The year 2019 was a record year for the cruise industry whole history. In 2020, plans for further development were thwarted by the COVID-19 pandemic, which almost completely ruined the global cruise ship industry. As the COVID-19 virus spreads around the world, large infection outbreaks have occurred on many cruise ships.

During the COIVD-19 each day that without operating cruise ships brought to this industry brought 110 million USD losses. It was estimated that globally, the industry lost 77 billion USD revenues in 2020 (www.businessinsider.com.pl).

This type of tourism in regular times can be recognized as the 'all-inclusive' mass tourism segment. Often, in ports of call (the cruise routes) passengers stay overnight on the ship and do not spend money on local hotels (De la Vina, Ford 1998). For the Caribbean Region tourists, who stay overnight on the islands, bring greater economic benefits than tourists, who take part in a cruise and stay on the island for one day or a few hours. Transfers abroad of profits from tourism are much larger for one-day visitors than for tourists. On ships, one cruise price usually covers all stay costs, which means that little tourist expenditure remains in the local economy. Moreover the major cruise ship companies are accused of hidden their revenues in tax havens and under-earning of their employees.

The number of the arrivals in Caribbean Region was 22.9 million in 2020, in 2005 it exceeded 25 million, in 2014 it exceeded 30 million and the record was before pandemic in 2019 with 37.4 million of arrivals (Fig. 1.4).

In the years 2000–2020, the share of cruise passengers to the selected island territories of the Caribbean Region was between 42 and 48%. In the year 2021, it

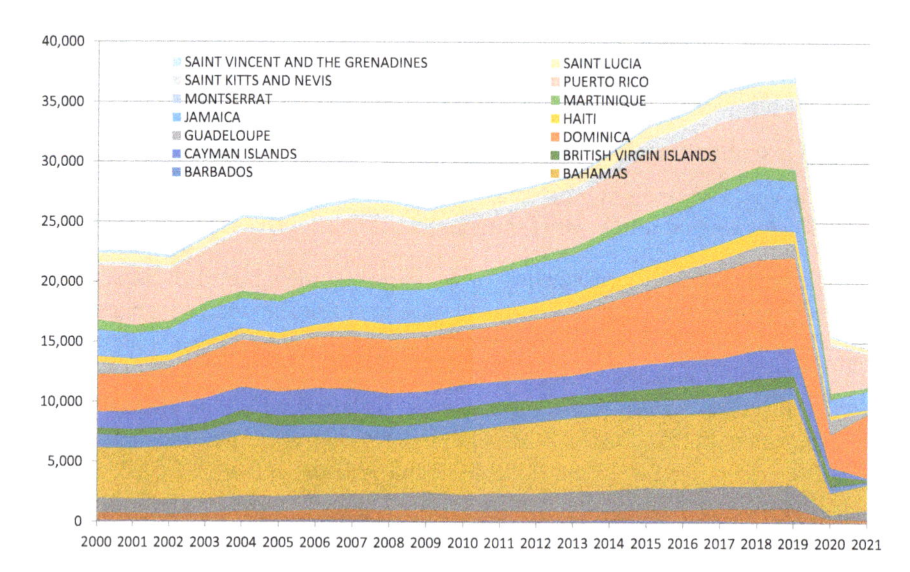

Fig. 1.4 Number of total arrivals to the Caribbean Region in 2000–2021 (thous). *Source* Own elaboration on the basis of UNWTO database

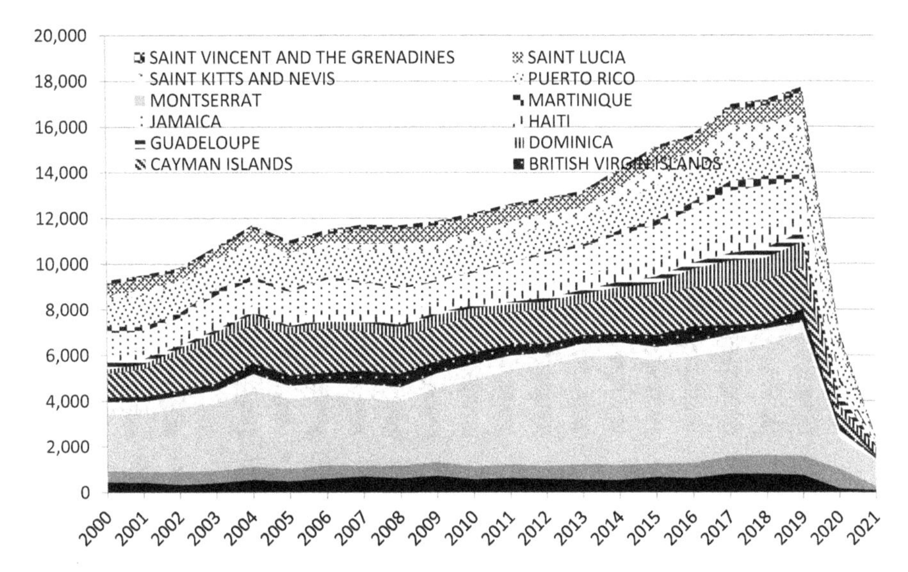

Fig. 1.5 Number of cruise-ships passenger's arrivals to the Caribbean Region in 2000–2021 (thous). *Source* Own elaboration on the basis of UNWTO database

was only 17% due to very difficult recovery of tourism sector during the COVID-19 pandemic. The cruise ships during the pandemic, where the places where the infection were on very high level. Many ports rejected to disembarkation of the ships, if they knew that there is a virus among passengers. So there was the icon of tourist problems for the people, who were stuck into the quarantine on such a ships during the first waves of COVID-19 virus.

Figure 1.5 shows the number of cruise-ships passengers arrivals to the Caribbean Region in 2000–2021. The best year for the cruise ship industry in the Caribbean region was 2019 with the almost 18 million of cruise ships passengers.

Before COVID-19 (in 2019) the highest rates of cruise passengers visiting islands in Caribbean Region had: Saint Kitts and Nevis (90%), Dominica (84%), Cayman Island (82%), Saint Vincent and Grenadines (82%), Bahamas (74%), Barbados (62%) and Anitua and Barbuda (55%),

The very interesting islands territory, which really suffers by mass tourism connected with cruise ship industry before COVID-19 pandemic was Cayman Islands. The number of arrivals exceeds almost 30-times the number of local citizens for these islands.

- **Cayman Islands**

The Cayman Islands are a British Overseas Territory with the surface of 259 km^2 and number of citizen's 80 thousands. The islands are located south of Cuba and they consists of three 3 islands: Grand Cayman, Little Cayman and Cayman Brac. Before COVID-19 in 2019 received more than 2.5 millions of visitors per year. The Cayman Islands have one of the highest rates of cruise passengers visiting the islands

Fig. 1.6 Number of total arrivals to the Cayman Islands (divided into overnight visitors and cruise passengers) in 2000–2020 (in thous). *Source* Own elaboration on the basis of UNWTO database

among all Caribbean Region 74% in 2020 till 91% in 2005, 82% in 2019 (average for 2000–2000 was 83%).

Tourism and financial services before the COVID-19 were the most important elements of the ¾ of population works in services. The value of GDP in purchasing power parity was 86 568 USD per citizen (in 2021), which is the one of the highest DGP per capita in the world. The Islands are known as a tax haven (lack of indirect taxes). Figure 1.6 shows the number of total arrivals to the Cayman Islands (divided into overnight visitors and cruise passengers) in 2000–2020.

The problems connected with overtourism phenomena in Cayman Islands are the inflation pressure. Many people are investing money in hotels and apartments in Cayman Islands and local are facing the problems with high inflation for the land and apartments value. Second problems is inflation connected with regular products, which are sold in the shops. The same problems are facing small tourist destinations in during tourism season.

According to Bromby (2021) Cayman Islands exhibit traits that 'may be described as insular or openly international'. However, the tourist flows shows many disadvantages connected with too much pressure on natural environment, social and economic sphere of the islands.

- **National parks and reserves in developing countries**

Conservation in the area of national parks and reserves located in less developed countries is a much greater challenge than in highly developed countries. These parks and reserves do not have large budgets to protect their resources, so their income must be based on tourism. If they are unable to attract enough tourists, their existence is at risk as local communities usually do not find arguments to continue protecting them. This results in overexploitation of the natural resources of parks and reserves by local people who hunt animals, set up agricultural plots, settle there and graze domestic animals within the boundaries of the protected areas.

On the other hand, the development of tourism, although profitable, may also be degrading for the nature of protected areas. Therefore, many planning concepts are used to limit the negative effects of the presence of tourists and related infrastructure. One of these concepts is zoning. It consists in designating zones with different intensity of tourist traffic or different types of this traffic, i.e., rock climbing tourists, speleologists, etc. By applying this concept, those areas that are particularly susceptible to destruction by tourism can be excluded from tourist movement.

Another concept is the dispersion concept, which consists in arranging tourist routes and attractions in such a way as to disperse tourists as much as possible in the protected area. It is very difficult to apply in practice because there are few protected areas with evenly distributed tourist attractions.

Another way is the concept of concentrating tourist movement in specially designated places, far from natural resources that could be destroyed by visitors. Places of this type are usually located somewhere close to the entrance to the park and have infrastructure designed to 'retain' tourists. It is in such places that there are toilets, tourist information, restaurants, educational paths, sports and recreation facilities, enclosures for the representatives of the animal world. Of course, some tourists will go deeper into the park anyway, but a significant percentage (i.e., families with young children or people with mobility difficulties) will stop in these places and will not feel the need to explore further parts of the park. An example of such '*honeypot*' in Monteverde Cloud Forest Reserve in Costa Rica are water containers for hummingbirds suspended near the entrance to the park, where visitors spend a lot of time trying to take a photo and observe these birds. Yet another concept is limiting access to a protected area through the introduced legal and administrative restrictions. Its aim is to reduce the number of tourists visiting protected areas. It is implemented through the limits of people or cars allowed into the park or reserve or, in the most popular form, through the admission ticket system.

- **Thilafushi—a trash island on Maldives**

The Maldives are extraordinary nature rich islands with astonishing coral reefs and beautiful environment. However, tourists are producing enormous number of waste and sewages. Thilafushi is an island called a 'rubbish island'. This is a place, where the huge amount of waste goes produced by over 100 thousand local residents and a lot of tourists, who are visiting Maldives. It is the fastest growing waste islands in the Maldives and it was introduced in 1992. On Fig. 1.7 were shown foreign tourist arrivals to Maldives in 2015–2019. We can observe almost 38% increase of the number of tourist arrivals to these islands during this period of time. Tourists are producing three times more trashes and garbage than the local residents.

- **Koh Phi Phi—Thailand**

This is overcrowded island because of tourism industry. It is also famous because of sandy beaches. It was devastated by tsunami in December 2004. The foreign tourist

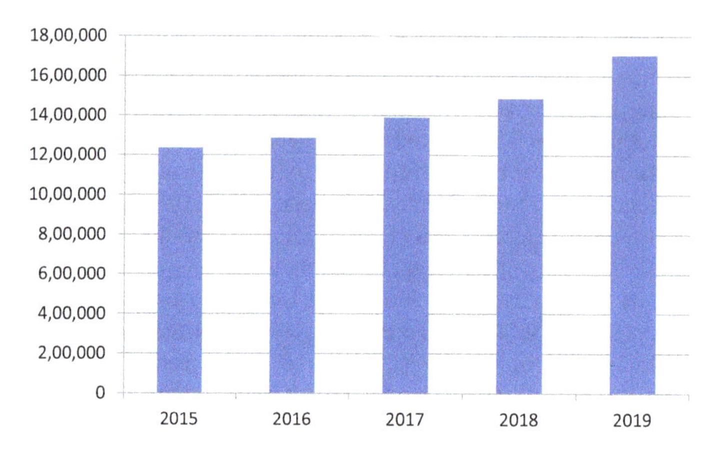

Fig. 1.7 Foreign tourist arrivals to Maldives in 2015–2019. *Source Yearbook of Tourism Statistics Data 2015–2019*, 2021, UNWTO, Madrid

arrivals in Thailand in years 2015 was 29.9 million and in 2019–39,9 million, which means 33% increase in foreign tourist arrivals (Yearbook of Tourism Statistics Data 2015–2019, UNWTO 2021). The island is administratively part of Krabi Province and accessible by speedboats or long-tail boats from Krabi. Initially was populated by Muslim fishermen. The number of local residents is about 2–3 thousands people.

On Fig. 1.8 were shown foreign tourist arrivals to Thailand in 2015–2019.

We can observe almost 33% increase of the number of tourist arrivals to these islands during this period of time.

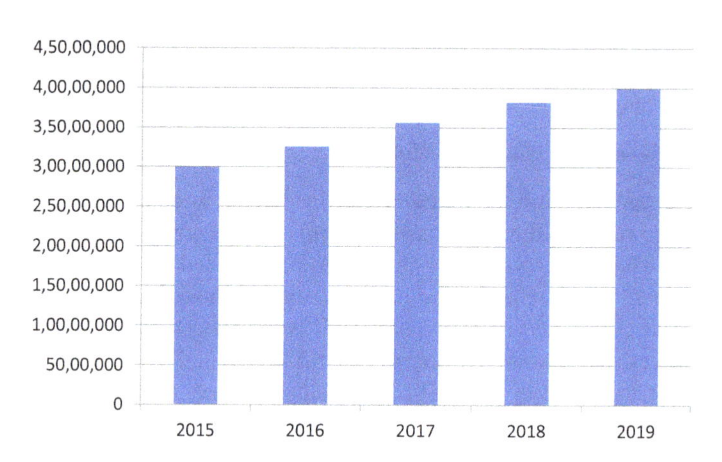

Fig. 1.8 Foreign tourist arrivals to Thailand in 2015–2019. *Source Yearbook of Tourism Statistics Data 2015–2019*, 2021, UNWTO, Madrid

1.6 The Sustainable Development Concept and the Example of Sustainable Tourism and Its Application in the Area of the Alps

This chapter deals with the sustainable development concept shown on the example of tourism sector the Alpine area, explains the concepts related to this type of tourism, presents examples of indicators for measuring the degree of sustainable tourism development, describes the strategic documents of the European Union and those adopted by individual Alpine countries in the field of tourism and sustainable development, as well as presents examples applying the principles of sustainable tourism in selected alpine regions—the High Alps in France, Graubünden in Switzerland and the Julian Alps in Slovenia. The chapter was written on the basis of the analysis of data from official databases of the European Union, content from scientific articles and publications included in the literature on the subject and the analysis of strategic documents at the international, national and local level. There are many sources on the problem of sustainable tourism, they are books, scientific articles and publications of government institutions. The period in which the problem of work is described covers the range from the second half of the twentieth century to the present day. The content of the chapter covers the territorial range of the Alps, with particular emphasis on the regions of the High Alps, Graubunden and Julian Alps.[1]

Tourism is a branch of the economy with a constantly growing importance in the global and European economy. The exception is, of course, the time of the COVID-19 pandemic, which has severely hit the tourism sector all around the world. As it is a relatively young sector, it is constantly evolving and changing. Currently, the world is changing at a pace incomparable to any previous period in the history of mankind. Therefore, from man and all areas of his life, an increasingly faster adaptation to the changing conditions of using the achievements of civilization and the resources of our planet is required. As a result, even from a relatively young tourism history, changes are required that will allow it to adapt to contemporary trends in the use of our planet's resources, which are characterized by greater respect for natural resources, caring for the environment and preventing its degradation. These are the features that tourism should have if it wants to meet the demands of conscious people from developed societies living in the twenty-first century. An ideal example of a tourist region that tries to implement the principles of sustainable development is one of the world's oldest tourist regions, which is the European mountain chain of the Alps. Located in the area of seven highly developed countries, it has undergone or is still undergoing transformation from a region focused on an outdated model of mass tourism, into a modernized model of sustainable tourism. The main reasons for this transformation are the potential economic, social and environmental benefits of this tourism model. Proper implementation of the concept of sustainable development in tourism will allow to maintain a large tourist movement in the region, despite

[1] This chapter was prepared on the basis of the bachelor degree of Filip Florczak defended in 2021 in Faculty of Geography and Regional Studies, University of Warsaw—Development of sustainable tourism and examples of applying its principles in the area of the Alps.

changes in the choices and behavior of European tourists, who constitute the vast majority of tourists in the Alps.

The aim of the chapter is to answer questions related to the development and state of sustainable tourism in the countries located within this mountain range. This study seeks for answers to the following questions: is tourism development in selected regions consistent with the principles of sustainable development? Are the examples of activities undertaken in tourism space discussed in the paper consistent with the principles of sustainable development? Obtaining answers to the above questions will help to achieve the research goal of the work. The validity of sustainable development concept the above theses will be examined on the basis of practical examples.

During the study of tourism indicators, an analysis of statistical data from verified, reliable sources such as Eurostat (the Statistical Office of the European Union), and the content of scientific articles published in reputable paper and online journals was used. Also analyzed were strategic documents at the international, national and local level, concerning the countries and regions discussed in the work—'Alpine Convention' and its protocols, 'Swiss Landscape' concept, 'Charter of good behavior', 'Strategy for the spatial development of Slovenia'. Based on the collected information and data, the model of sustainable tourism was characterized and its impact on the tourism sector of the described countries and regions.

The territorial range of the area in question is, in a broader context, the borders of the Alpine countries (i.e., those whose whole or part is located in the Alpine area). These include the following countries: France, Switzerland, Italy, Liechtenstein, Austria, Germany and Slovenia. The regions of these countries located directly in the Alps will be analyzed in detail, examples from three selected regions will be discussed: the High Alps in France, Graubünden in Switzerland and the Julian Alps in Slovenia. Looking at the situation in the region as a whole and citing specific examples of places, it will be verified whether these regions act in the development of tourism on the basis of sustainable development. The time range discussed in the text covers a period of about 70 years, from the beginning of the second half of the twentieth century until the present day.

1.6.1 The Concept of Sustainable Tourism and the State of Its Implementation in the Alps

Sustainable tourism is a type of tourism that is becoming more and more popular among decision-making bodies and the local population, which can be seen when looking at strategies being developed and measures taken to develop it in an increasing number of countries. In order to better understand the concept of innovative, modern sustainable tourism, it is first necessary to look at the concept of tourism and the idea of sustainable development. Tourism is a multidimensional phenomenon bordering on economics, geography and sociology. It is a social, spatial and economic phenomenon, and its development depends on many factors, external conditions and human behavior (Niezgoda et al. 2020). Sustainable development is intergenerational solidarity, which consists in applying solutions that guarantee further growth, enabling active inclusion of all social groups in development processes, while giving them the opportunity to benefit from economic growth (Niezgoda et al. 2020). Important factor in sustainble tourism development is waste managment in hotels (Bugdol and Puciato, 2023).

When trying to develop mountain areas, attention should be paid to several important issues regarding the appropriate distribution of tourist infrastructure. These include, among others, ensuring appropriate conditions for reaching the tourist area, preparing an appropriate accommodation and catering base, paying attention to climatic and weather restrictions, building the appropriate technical infrastructure allowing for the free use of mountain areas, as well as preparing an entertainment, cultural and cultural offer. Sports available for tourists in the off-season and independent of weather conditions (Kowalczyk and Derek 2010).

Many socio-economic benefits of the idea of sustainable development have been noticed in the European Union, which is why it has recently become willingly promoted by Member States, and its assumptions are more and more often included in strategic documents such as: 'Europe 2020—a strategy for smart and sustainable conducive to social inclusion'.

The development of this concept in the field of tourism is sustainable tourism, which is a concept of eco-development aimed at minimizing the negative impact of tourism on the natural environment and local culture, while generating jobs for local residents. Much emphasis is placed on the integration of nature protection objectives with tourist activity, including the development of new, ethically and socially beneficial for the local population, behaviors and attitudes among tourists and people organizing tourism. To meet these assumptions, it is necessary to apply the following principles:

- adapting the development of tourism to the natural resources of the environment in a way that does not contribute to their degradation,
- participation of the population in all activities related to tourism, undertaken in the areas inhabited by it,
- tourist facilities should be of a small scale and their offer should be based on local resources,
- there should be an integration of tourism development and local economic development, and the development itself should take place in some ethical, social and economic benefits for the local population.

It should also be remembered that sustainable tourism does not mean the same as terms such as ecotourism, nature tourism or agritourism. In contrast, sustainable tourism is not a specific form of tourism, but a more universal term for any form of tourism that meets the goals and theoretical assumptions ascribed to sustainable tourism (Niezgoda et al. 2020). According to Kamieniecka (1995), ecotourism is 'practicing selected forms of tourism closely related to the use of a specific element selected from natural resources or tourist Values.' According to many researchers, nature tourism is not recommended to be used interchangeably with the term sustainable tourism, because caring for nature protection cannot be a brake on socio-economic development (Kowalczyk Derek 2010).

Highly developed countries, which include, inter alia, the Alpine countries, have readily adopted and widely applied the idea of sustainable tourism. By focusing on active tourism aimed at small groups with high mobility, which is to promote the culture and traditions of holiday resorts, they want to reverse the devastation in local culture and business, which was caused in popular tourist regions by the spontaneous, dynamic development of mass tourism in the second half of the twentieth century (Gołembski 2009).

It is also worth mentioning the great importance of the natural environment in the development of sustainable tourism. By ensuring the value and attractiveness of the tourist product, it takes on the negative effects of tourism and its development to the greatest extent. While taking care of tourism as a priority function, we cannot forget about the sustainable development of other sectors. To this end, appropriate plans and strategies should be prepared to support other sectors and to avoid future conflicts caused by the neglect of the interests of other groups related to the tourism sector. When implementing plans and strategies, they are required to closely cooperate in long-term awareness-raising activities about the role of tourism and an ecological approach that is safer for local culture and the natural environment.

A well-conducted tourism policy in the environmentally valuable mountain areas of the Alps requires understanding the tourist-environment relationship well in advance, when there is no threat caused by their uncontrolled interaction, as well as highlighting the strengths of specific tourist regions in the form of uniqueness and authenticity produced by them of the product (Niezgoda et al. 2020).

1.6.2 Sustainable Tourism Indicators

Indicators are variables that are subject to measurement and remain in a constant relationship with the analyzed features of states of affairs or objects. According to the European Commission, they can also be treated as a measure in which we measure the degree of achievement of the goal, the number of resources involved, the effect to be achieved or the quality of a given phenomenon (Górniak 2008).

In 2013, the European Commission developed a pilot set of sustainable tourism indicators, 'The European Tourism Indicator System toolkit for sustainable destination management' abbreviated as ETIS. It has been designed for the local level to monitor sustainable tourism management processes and facilitate the sharing of their results in the future. However, it is not a rigid set that would require the municipalities or localities applying it to use all the indicators provided in it, but rather a set from which interested parties can choose an individual set of indicators tailored to the situation in the municipality or town. These indicators are determined by two criteria:

- **Usefulness**—functionality and usefulness for people managing tourism development so that they meet the assumptions of the concept of sustainable development,
- **Accessibility**—accessibility for managers, there is a rule that the cheaper and easier it is to obtain data, the greater their availability.

Proper evaluation of sustainable tourism requires actions in specific areas to maximize the effectiveness of actions aimed at the protection of natural resources and biodiversity (Niezgoda et al. 2020). The proper development of sustainable tourism consists of many factors and sectors, requiring a large amount of work from the local community in many fields of activity, but with appropriate guidance and commitment, there is a great chance of success and bringing the expected benefits.

The development of sustainable tourism in highly developed Alpine countries can be observed in the analysis of indicators presenting statistics on the state of European tourism in recent years. Below it was presented a few of them, where in some of them and showed in particular at how they are presented in the Alpine countries of France, Switzerland and Slovenia.

The study of indicators used data from Eurostat, which is a reliable database with up-to-date data from recent years. This allows for an insight into statistical data from the last dozen or so years at the level of selected regions at the NUTS 2 level, where the analysis may include indicators showing the aspect of tourism in the form of the impact of tourism on the economy, the presence of the phenomenon of second homes, the number of beds in rural areas, the share of tourists foreign, seasonality of tourist movement and the means of transport used by travelers.

In Table 1.3 were presented the analyzed indicators of sustainable tourism.

The tourism revenue-to-GDP ratio in 2018 was the highest in Slovenia, where it reached 5.9% of gross domestic product, in Switzerland it was 3.9% and in France it was 2.4%. We can notice a tendency here that usually the smaller the country in terms of population, the greater the share of tourism in the entire economy of the country. The indicator shows the importance of tourism in relation to other sectors of the economy, and its high value suggests a high degree of economic exposure, caused by an excessive dependence of general economic development on the tourism sector (Niezgoda et al. 2020).

Table 1.3 Analyzed indicators of sustainable tourism

Analyzed indicators
– Ratio of revenues from tourism in relation to GDP (in %) – The phenomenon of second homes – Number of beds per square kilometer in rural areas – Share of foreign tourists (in %) – Seasonality of tourist movement – Use of particular means of transport when EU-27 residents travel to other EU countries

Source Own elaboration on basis Eurostat Database, European Commission

It was worth mentioning the phenomenon of second homes, popular in this region of Europe, consisting in the rental or purchase of real estate located in popular tourist regions or attractive in terms of nature by representatives of the middle class and higher developed countries, mainly for recreational purposes. In 2012, the number of such properties in the Alps was estimated at almost 2 million, representing 26% of the region's total housing stock. This phenomenon mostly affects the western regions of France and Italy, which are developed in terms of tourist infrastructure. Eastern regions in Slovenia are the least popular for second homes, but over the past two decades there has been a significant increase in their number in the region (in 2012, it was around 40,000 apartments) (Koderman and Pavlič 2019). This indicator shows the state of investment development that affects the natural environment, but does not bring such benefits as traditional accommodation (Niezgoda et al. 2020).

When examining the indicator of the number of beds per square kilometer in rural areas, we can observe that, compared to the rest of Europe, the Alps are distinguished by a high number of such places. For example, in western Austria, northeastern Italy and the Italian northwestern Aosta Valley, the index is at its highest, ranging from 14 to 35 beds per square kilometer. It is related to the history of these regions, in which the rural population from mountainous areas did not leave their native regions en masse, and therefore there is a higher share of private owners of accommodation units scattered in smaller towns, and the share of large centers focused on mass tourism located in the area of cities is smaller. On the one hand, this leads to a larger area of potential environmental degradation, but on the other hand, the degree of degradation is distributed to a greater extent, which reduces their intensity, and thus their effects (European Environment Agency 2017). Figure 1.9 shows the number of bed places per square km in rural areas at the NUTS 2 level in the European Union in 2017.

When it comes to the share of foreign tourists in the Alpine regions, it can be noticed that a higher share of tourists is observed in the eastern part of the Alps (in 2018, in Tyrol and Liechtenstein, the share of nights spent by foreign tourists in the total number of nights spent by tourists is over 90%), while the lowest in the western part of the Alps (in 2018, in the French regions of the Rhône Alpes and Provence-Alpes-Côte d'Azur, it is approximately 30%) (European Commission 2018). For Slovenia, the share of foreign tourists in 2019 was 72% nationwide. We can see the following relationship between the size of the country and the share of foreign tourists—the greater it is, the smaller the share. As the western part of the Alps is located in large countries (France and Italy) and the region is popular among native tourists, there is a relatively low share of foreign tourists there (Slovenian Tourist Board 2021). Table 1.4 presents the total number of overnight stays together with the percentage share of foreign tourists in the total number of overnight stays in selected regions.

When examining the seasonality of tourist movement in the European Union, we can observe that the most popular months when it comes to tourist trips are July and August (only in these two months there were almost a quarter of the total number of trips of EU residents). There is also a higher number of long trips (4 nights or more) during the summer months (*Eurostat Statistics Explained* 2019). This is due to the

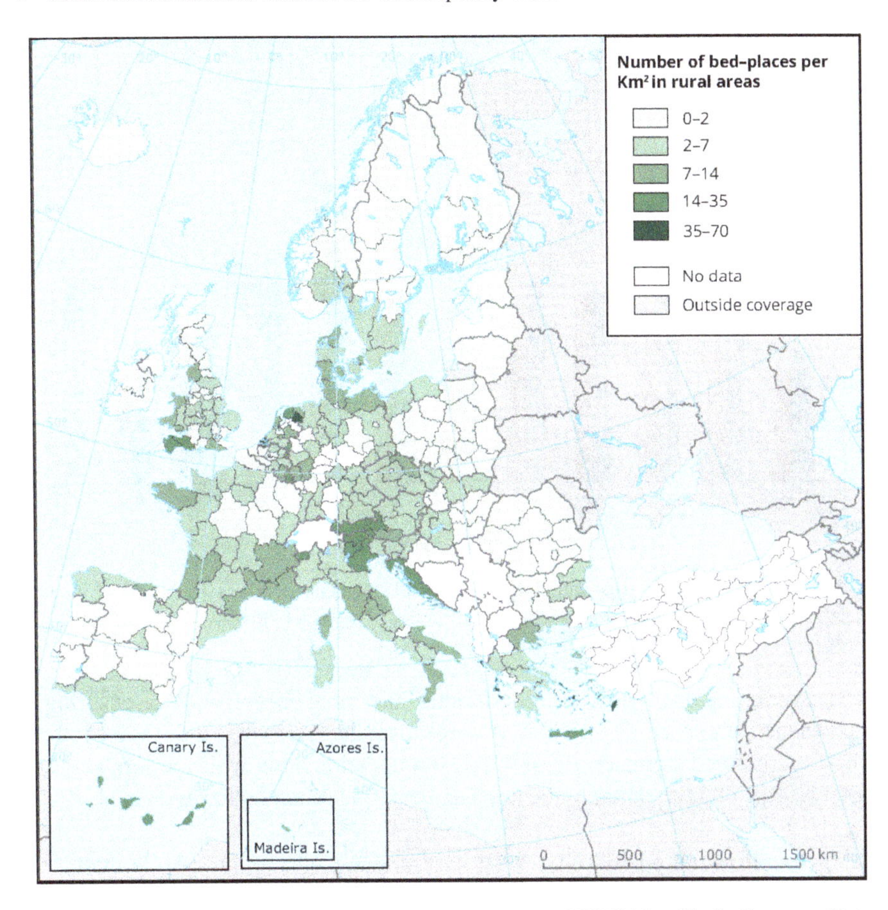

Fig. 1.9 Number of beds per square kilometer in rural areas at NUTS 2 level in the European Union in 2017. *Source Number of bed places per square in rural areas*, (2017), European Environment Agency

Table 1.4 Total number of overnight stays and the percentage share of foreign tourists in the total number of overnight stays in selected NUTS 2 regions

Region NUTS 2 (country)	Total number of nights (in millions)	Share of foreigners in the total number of nights (in %)
Provence-Alpes-Côte d'Azur (France)	54.8	32.2
Rhône Alps (France)	50.9	29.9
Tyrol (Austria)	30.8	90,6
Liechtenstein (Liechtenstein)	0.2	97,9
Slovenia (Slovenia)	15.7	72

Source Own elaboration on basis: Regions in Europe—Statistics visualized, European Commission

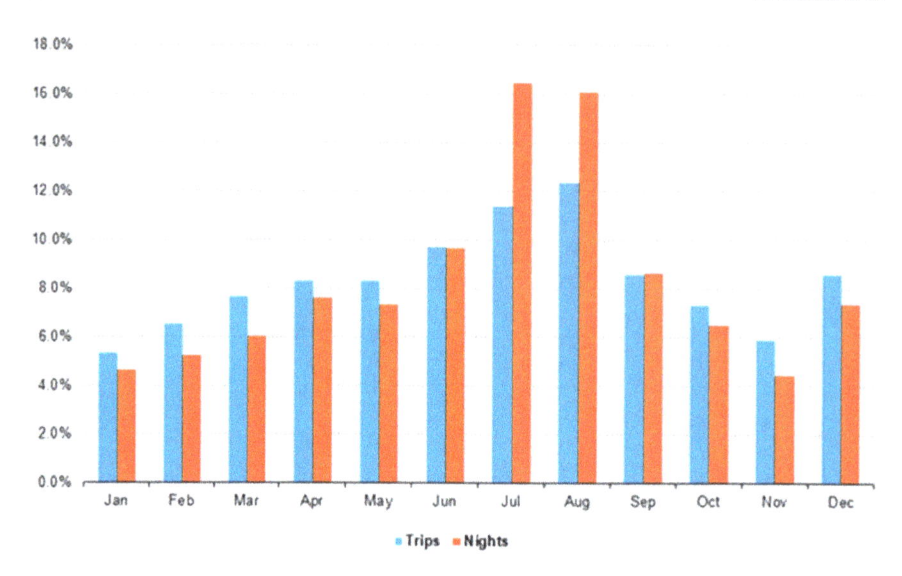

Fig. 1.10 Monthly share of trips and nights spent by EU residents in 2017 (% share of 12 months in total). *Source Seasonality in tourism demand,* Eurostat Statistics Explained

higher average temperatures in these months and more secure weather ensuring a better chance of a successful holiday. A longer average tourist stay reduces the inflow of new tourists and a greater use of the accommodation base, which in turn leads to higher returns on investment in the hotel and catering sectors (Niezgoda et al. 2020) (Fig. 1.10).

Among all means of transport used by the inhabitants of the EU-27 countries when traveling to other EU countries, in 2018 air flights (43%) and motor vehicle journeys (42%) prevailed. Other, less popular means of transport are buses (6%), railways (4%) and waterways (3%) (European Commission 2019). This proves the prosperity and development of this part of the world, whose inhabitants can afford frequent flights by airplanes and long distances using private vehicles. The indicator also shows how transport contributes to the increase in the level of local air pollution (a large share of motor vehicles), which may prove useful in the development of appropriate transport strategies that will allow for the proper development of tourism in spatial terms and the promotion of optimal methods by managers arrival (Niezgoda et al. 2020). Table 1.5 presents the share of individual means of transport used by residents of the EU-27 countries when traveling to other EU countries in 2018.

In Table 1.6 were shown the general trends of indicators by country size. Larger countries with diversified economy have the relation of revenues from tourism to GDP smaller than the countries with small and not so diversified economy.

The conclusions that can be drawn from the analysis of the indicators are as follows:

- The Alps are a region with visible socio-economic differences between individual countries, caused both by different country sizes, geographical location and historical conditions,

Table 1.5 Share of individual means of transport used by residents of the EU-27 countries when traveling to other EU countries in 2018 (%)

The country of destination	Airplane	Water	Train	Coaches	Motor vehicles	Other
EU-27	43	3	4	6	42	1
Germany	34	1	7	8	50	1
France	28	1	8	6	54	3
Italy	41	2	3	9	44	1
Austria	11	< 1	7	7	73	2
Switzerland	23	< 1	14	5	56	< 1
Slovenia	20	< 1	< 1	9	68	< 1

Source Own elaboration on basis: Eurostat Statistics Explained, European Commission

Table 1.6 A summary of the general trends of some indicators in the form of a table broken down by country size

Indicator	Larger countries (France, Italy)	Smaller countries (Switzerland, Austria, Slovenia)
Revenues from tourism to GDP	Smaller	Bigger
Share of foreign tourists	Smaller	Bigger
The magnitude of the phenomenon of 'second homes'	Bigger	Smaller
Share of motor vehicles in total means of transport	Smaller	Bigger

Source Own elaboration

- when examining the above data, the image of the Alps becomes visible as a region of Europe rich, but at the same time well-developed in terms of infrastructure in rural areas,
- the region is widely appreciated abroad, as evidenced by the high number of tourists (including a significant share of foreign tourists), as well as the popularity of the 'second home' phenomenon.

Examples of actions taken at the international and national level.

Changes in the approach to the development of the Alpine region were inevitable, and the Alpine states noticed this long ago. In order to ensure better inter-state cooperation, in 1952 a non-profit panalpine non-governmental network called 'CIPRA' was established, bringing together representatives of 7 Alpine countries and hundreds of smaller organizations and institutions. The activities of this organization led to the signing of the 'Alpine Convention' in 1991 by the European Union and the Alpine countries, making it a legally binding document. This document ensured that actions were taken on border cooperation between individual countries, regions and local communities. The convention shows how important a factor in sustainable development is the cultural landscape (Mrak et al. 2012).

As part of it, several protocols have been drawn up on the subject of environmental protection and the pursuit of sustainable development in the Alps. The 'Mountain Farming Protocol' of 1994 deals with measures to protect mountain farming that will be tailored to the individual situation in the region and will be safe for the environment. It also draws attention to the important role of agriculture in maintaining traditional settlements and sustainable economic activity, which in turn ensure high-quality products, protect the natural living space and help maintain local natural and cultural values. In 1996, the 'Mountain Forest Protocol' was adopted to protect and improve the stability of mountain forests as a sustainable living space. After them, in 1998, the 'Tourism Protocol' was created, which contributed to the sustainable development of the alpine space with tourism activities acceptable to the space of these mountains. In order for this development to be achieved, two conditions must be met:—prioritizing a balanced tourist offer that focuses on quality over mass tourism, and taking care of local agriculture, creating the cultural landscape, which is an important factor contributing to biotic diversity (Mrak et al. 2012).

Actions for the application of the concept of sustainable development are one of the most important priorities of the spatial policy of the European Union. As a result, many strategic documents were created that allowed the EU countries to guide this matter. The European Spatial Development Perspective (in 1999), the European Landscape Convention (in 2000) or the European Sustainable Development Strategy (in 2001). The documents indicate the values of maintaining harmony in the landscape and the natural environment, such as social welfare, benefits for economic activity or creating the value of society (Myga-Piątek 2011) (Fig. 1.11).

In 2001–2014, the European Commission published several communications which presented political guidelines for the further development of the tourism sector, presented sustainable development as a recipe for the long-term competitiveness of tourism, an analysis of its factors and obstacles to its implementation, and a plan to increase the number of tourist visits from third countries (*Fact Sheets on the European Union 2021*, 2021).

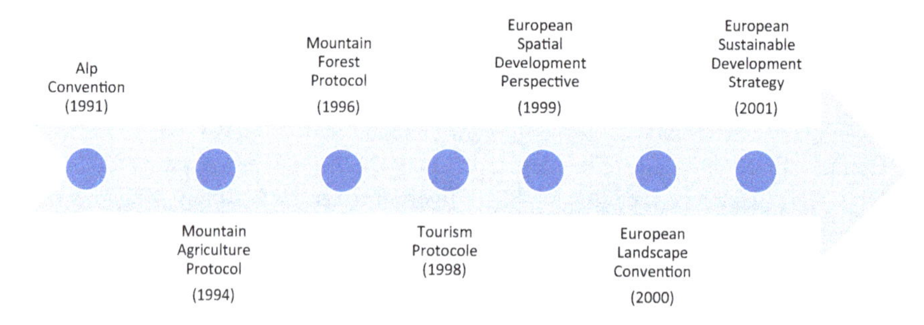

Fig. 1.11 Presentation on the timeline of the most important documents concerning the development of the concept of sustainable development in Europe in the years 1991–2001. *Source* Own elaboration

The European Union, when preparing special preparatory programs of issues important for European tourism, places emphasis on the issues of sustainable development. Initiatives such as 'Eden'—promoting little known tourist destinations operating according to the principles of sustainable development, or 'Sustainable Tourism', which covers the 6.800 km long stretch from the coast of the Barents Sea to the Black Sea along the former 'Iron Curtain'—encouraging visits less popular regions of Central and Eastern Europe can be examples here (*Information papers on the European Union*—2021, 2021).

In 2018, the EU-co-financed COSME cross-border program was launched, which aims to diversify the European tourism offer. As part of it, the regions concerned can apply for support in the development and promotion of tourism products, using synergies between tourism and the cultural and creative sectors. The program launched actions such as:

- supporting growth in the tourism sector to make it more sustainable, but also competitive,
- supporting products and services from the sport and wellness sector and cultural heritage,
- facilitation for younger and older EU citizens during tourist flows across borders in periods with less tourist movement in the off-season,
- striving for the greatest possible synergy between tourism and the cultural and creative industries (*Information documents on the European Union—2021*, 2021).

Examples of actions that have been taken in the field of sustainable development and sustainable tourism at the state level in the Alpine countries of France, Switzerland and Slovenia are presented below.

Since the 1970s, the French authorities at the governmental level have taken a number of decisions aimed at changing the way of spatial development in mountain villages to one that will be less harmful to the environment. The most important of these were the exclusion of areas with natural nature from tourism development plans, limiting the size of the space occupied by sports facilities and recreation areas, a decision on the obligatory opinion of expert committees for new projects, or the issuance of an act of decentralization of investments, giving more scope for local communities to act, while limiting the role of the state (Kurek 2001).

The French Ministry of Tourism at the end of the twentieth century developed a special plan to improve the standards of tourism in mountain areas. It concerned not only sports centers, but also other centers with other functions. The most important assumptions of this plan in terms of sustainable development were:

- improving communication in the centers by extending pedestrian roads, limiting car traffic and developing public transport,
- reorganization of the operation of services and their opening hours,
- undertaking activities aimed at revitalizing the historical parts of the village,
- introduction of recreational methods alternative to skiing in order to become independent from weather conditions (Kurek 2004).

In Switzerland, the state authorities also realized how big a threat to the environment and local culture is the uncontrolled development of tourism. Therefore, documents aimed at drawing attention to the current situation have been adopted. An example is the Conception 'Paysage Suisse' introduced in 1997, from which you can learn that the landscape, architectural and landscape values constitute the national heritage and require an appropriate pro-ecological tourism policy of the state. Another important document is the so-called Charter of good behavior (French 'Charte comportementale') which is a recommendation for citizens to respect the environment by respecting plants and animals, stop littering the visited places or use more economical means of transport such as bicycles or public transport, while limiting the use of car (Klimek 2010).

These activities resulted in the fact that at present Switzerland has one of the most environmentally conscious societies in the world and is one of the leaders in engaging in international activities in the field of sustainable development. In 2007, the Swiss authorities arranged a second conference on climate change and tourism organized in Davos by the World Tourism Organization. The Davos Declaration was signed there, the aim of which was to sensitize countries and entities operating in tourism to climate change (Klimek 2010).

In Slovenia, in 2004, the 'Strategy of Spatial Development of Slovenia; was adopted, which states that for mountain areas (in which the area of the Slovenian Alps is completely located) tourism and agriculture should be environmentally oriented and use renewable sources as much as possible. Mountain regions are usually more active in these aspects and, as a rule, show greater possibilities of their implementation with adequate support provided to communities living in mountain and hilly areas (Mrak et al. 2012).

1.6.3 Selected Case Studies

The development of tourism in the spirit of the concept of sustainable tourism will be analyzed on the basis of examples from three selected regions located in the Alps in three different countries. With the example of each region, first will be presented the current state of alpine tourism in the area of the country where it is located, emphasizing the previously discussed categories of tourism, such as skiing and summer tourism. Then an example will be shown of an action aimed at developing sustainable tourism in a given mountain region, discussing the regional conditions and what effects the action has had for the region.

- **High Alps Region (France)—an example of innovative activities in a commune**

Used in the French Alps in the 60 s-70 s. In the 1980s, the model of tourism development, which was based on the use of 'growth centers', which were ski resorts, caused many negative effects on the natural environment, such as air pollution, excessive water abstraction, and degradation of soil, vegetation and the landscape as a whole.

The so-called the Snow Plan developed by the French authorities, consisting in the simultaneous construction of about 20 ski resorts with a capacity of 150 thousand accommodation places can be an example of such activities. Despite the large economic benefits related to the development of winter tourism caused by the increase in tourist movement and the provision of new jobs, the specter of negative ecological effects began to appear over the region, which over time began to manifest in a series of catastrophes, the most common of which were avalanches and snowstorms, cutting off some ski resorts from the world. For this reason, in the mid-1970s, such a policy of developing winter stations was abandoned, focusing on more sustainable development (Konopinska 2017).

Currently, in France, mountain tourism develops throughout the year, in summer it is hiking and mountain climbing, while in winter, skiing is the most popular. In the French Alps alone, there are 218 ski stations, and the area for skiing there is about 118 thousand ha, which is much higher than the rest of the Western European Alpine countries. Hiking and climbing are practiced annually by almost 4 million tourists, and the most important mountain hostel operator, the French Alpine Club (CAF), has 91 facilities in the Alps. It is also worth mentioning rural tourism, which accounts for about 1/3 of the total tourist movement of the French. Such a large development was possible thanks to the great importance of the second home phenomenon, rich family and social relations, and the great importance of culinary traditions. In the characteristics of regional tourism, the concept of green France (French verte) is distinguished, which is defined as rural, mountain and forest areas of the country in the areas of villages and rural settlements (Kurek 2012).

The French Alps, which are small in size, are characterized by seven ski regions. The most famous ski station is Chamonix, lying at the foot of Mont Blanc, which is also considered the world capital of mountaineering. Another popular destination is Espace Killy, located between Val d'Isere and Tignes on two large glaciers, with 300 km of ski slopes with the best conditions at the beginning of the ski season and one of the best areas in the world for off-piste skiing (Jędrusik 2010).

French ski resorts, including those located in the High Alps, usually stand out from those in other Alpine countries in terms of their larger size, assets and turnover. Therefore, they also have higher profitability due to greater use of economies of scale and greater marketing resources (however, this is closely related to their size, not the fact that they are located in France, in fact, ski resorts in this country usually do worse than, for example, in France. in Austria). Such large centers are less vulnerable to the negative effects of climate change due to a lower level of debt giving greater financial autonomy, which in turn translates into the possibility of taking risks more willingly in the form of new investments without fear of risking financial imbalance. Due to the presence of such large centers, French tourist municipalities are more exposed than others to economic dependence on the service sector, which could be a threat in the medium term due to the shortened duration of the ski season. These municipalities are also vulnerable to tourism conflicts with core activities, which is a big problem for sustainable development. In order to avoid these problems, a complementary relationship should be established between them, preventing damage to the primary sector in the medium term, loss of natural resources and landscape degradation by tourism activities (Moreno-Gené et al. 2018). Actions are also being taken to change the character of French mountain villages in the High Alps region. Instead of being just a background for powerful resorts, towns and villages try to become independent by introducing a diversified, year-round offer that reduces economic dependence on large enterprises.

An example of innovative activities in the field of sustainable development in the region is the case of the Pays des Ecrins region and the commune of Argentière-la-Bessée, which was chosen to host the Alpine Week in 2008 because of the changes it has experienced over the last 20 years toward innovation in the management of mountain areas (Bourdeau 2009). Figure 1.9 shows the location of the Pays des Ecrins region and the Hautes-Alpes Department on the Provence-Alpes-Côte d'Azur map and on the map of France (Fig. 1.12).

Until the 1980s, the commune was characterized by bipolarity between large-scale industrial plants and smaller-scale traditional activities, such as agriculture

Fig. 1.12 Location of the Pays des Ecrins region (dark red) and the Hautes-Alpes Department (light red) on the map of the Provence-Alpes-Côte d'Azur Region and on the map of France. *Source* www.commons.wikimedia.org/wiki/File:Gemeindeverband_Pays_des_%C3%89crins_2018.png (09.06.2021)

and tourism. Once a municipality with a developed heavy industry began to deal with an economic, demographic and political crisis after the closure of Pachiney's largest aluminum factory in 1986, which resulted in the loss of over 300 jobs by the local population and a significant reduction in the municipality's population by 15%, and also resulted in the disintegration of the city council. Even the 70-million-franc industrial conversion plan financed by Pachiney with the support of the public authorities did not help the community. In the face of such a great catastrophe, voices about the inevitable changes in the development of the commune could be heard louder and louder. It was decided to choose a new road by the commune, (i.e., to promote tourism and activities related to cultural heritage, as well as to pay more attention to the development of mountain areas). With the revised 'rejuvenated' council determined to achieve this goal and new local entrepreneurs—guides, hoteliers, and shopkeepers—concrete strategies began to be developed to achieve these goals. Therefore, in the early 1990s, a multiannual tourism development plan was launched, which was intended to use the rich, unused natural and cultural resources offered by the Durance and Fournel valleys (Bourdeau 2009).

Starting from the 1990s, the change in the focus of the region was manifested in activities and indicators such as:

- Creation of a scientific, technological and industrial culture center (CCSTI),
- Expanding the network of climbing paths,
- Organization of sports events at an international level,
- The emergence of numerous professional tour operators,
- Establishing a greater number of trail markings and information boards,
- Membership in the network 'Plus beaux detours de France' ('The most beautiful detours in France').

Summer sports tourism started to have a great impact on the development of the region, which allowed for a completely new shaping of space, different from that in the case of traditional activities related to recreation in the mountains. This allowed for the development of previously neglected areas such as the bottom of the valley, villages or the areas around the Puy-Saint-Vincent resort. Sports tourism changed its character from seasonal, intended for a longer period of free time, such as holidays or holidays, to weekend, everyday character, where the priority when choosing an activity was changed from the environment in which it is performed to the movement of the body itself (Bourdeau 2009).

This region is very flexible and has the ability to initiate or accept experiments. Examples could be here:

- Creation of concept via *ferraty* and the introduction of the first French route for this activity,
- the signing of the first Natura 2000 convention between the government and local authorities,
- Carrying out the first voluntary SCOT project in France and creating a sustainable development plan,

- Performing an experimental environmental analysis in the Pelvoux-Vallouise resort.

In addition to referring to the concept of sustainable development, the creativity observed in the region was accompanied by reflection on the observation of change, but also an understanding of the meaning of action and the best use of the experience gained (Bourdeau 2009).

The following meetings and activities helped this commune:

- Conference 'Natura 2000 and local authorities' (2002)
- Dynalp2 Seminar (2007),
- Meetings on transalpine routes (2008),
- Alpine Week (2008),
- Support for French and foreign universities,
- Numerous analyzes and consulting research.

All this means that today the commune serves as an example of a region that has undergone modernization from an obsolete, industry-oriented region, to a region with developed modern tourism, with an idea for itself, ready to meet the challenges posed by the change in tourist habits of Europeans and the inevitable, progressive climate change.

- **Graubünden region (Switzerland)—an example of synergy between agriculture and tourism**

The Alps cover about 60% of Switzerland's surface, and the permanent glaciers of the highest parts of the mountains cover about 3000 km^2. The Swiss Alps have always been characterized by being an international tourist area, where, unlike the neighboring Alpine countries, there was no clear advantage for tourists from one country. The summer season, characterized by a high proportion of hiking, runs from June to September (in some places from May to October). Therefore, the country has a highly developed infrastructure in the form of numerous cableways and a dense network of marked tourist routes with a length of more than 50, 000 km, including three types: valley trails—Wanderwege, mountain trails—Berwege and alpine trails (Kurek 2004).

As Switzerland is characterized by a large number of people skiing (2.4 million inhabitants in 1998), the country also has a developed winter sports infrastructure. There are several hundred ski resorts in the country, which are mainly characterized by their small size and location in the lower parts of the mountains, often within the countryside, where they are jointly owned by the locals. This makes these stations consistent with the spirit of sustainable development, as they are more environmentally friendly and provide income for the local population. There is also a tendency to decrease the number of ski lifts in the country. In the years 1979–1998 their number decreased from 1869 to 1757, but thanks to the appropriate modernization, the total capacity increased by about 3%. In Switzerland, approximately 11.000 people work in the companies operating lifts and cableways alone and other thousands are employed in services related to skiing (Kurek 2004).

Fig. 1.13 Location of the Grisons canton (red color) on the map of Switzerland. *Source* www.commons.wikimedia.org/wiki/File:Kanton_Graub%C3%BCnden_in_Switzerland.svg (09.06.2021)

It is estimated that every tenth resident of Switzerland works in the tourism sector and industries directly related to it. Therefore, there is a systematic increase in the importance of tourism, with a simultaneous decline in the importance of agriculture (Klimek 2010). This is well illustrated by the trend in the decline in the number of farms in recent years. The decrease in their number by half and the progressive emigration from rural areas, with the simultaneous development of technology giving higher yields and work efficiency, contributed to the dynamic development of agritourism since the 1960s (Kurek 2004).

Graubünden is the most popular tourist canton in Switzerland, where, according to data from 2003, about 5.6 million tourists spent the night. In 2007, tourism generated 30% of local GDP and jobs in this region (Klimek 2010). There are about 46 ski areas in Grisons (Kurek 2004) (Fig. 1.13).

Graubünden has great ability to combine the potential of organic farming with tourism. This is due to the fact that every second farm in the canton is an organic farm, and almost 3.6% of people working in Grisons find employment in this type of farm. Additionally, since 2011 there has been an increase in the number of organic farms in favor of conventional farms. This is perfectly illustrated by the example of a project implemented in catona aimed at optimizing the strengthening of cooperation between the organic farming and tourism sectors, which is to stimulate the sustainable development of the region and increase the efficiency of the organic farming segment (Kuhnhenn and Simon 2013).

The development of the project was to be a response to the change in trends in mountain tourism in favor of summer tourism due to structural and climatic transformations, as well as a reorientation in travel motives to those containing ecological and social values, having unique and authentic features, which can be found in natural, regional products. An example of such tourism can be the 'natural tourism' already existing in countries such as Austria or Switzerland, which offers tourists experiences related to intact natural and cultural landscapes. The second impetus for the project came from changes in the agricultural sector, where farm owners began to develop new sources of income. The best solution turned out to be rural tourism, which was to be an advertisement encouraging people to visit from outside. There have also been changes in production due to the increasing share of organic farming. All these changes have brought both sectors of the economy closer to each other (Kuhnhenn and Simon 2013).

As one of the most visited regions by tourists and a canton with a high share of organic farming, Graubünden has ideal conditions for such cooperation and optimization of operations for the entire value chain. In doing so, it realizes the following sub-goals:

- develops innovative logistics and sales concepts, which guarantees the quality of ecological products that can be easily delivered in an efficient and cost-effective manner,
- develop concrete proposals for an innovative approach to cross-sector products and a service portfolio.

To achieve them, the project is carried out in four phases, of which the first phase has already taken place:

- preliminary research,
- survey and analysis,
- concept and implementation,
- evaluation.

After the preliminary survey, the respondents from the region expressed the opinion that the adopted policy of reaching to sources should be continued in order to secure additional income, and the best way to achieve this is to expand the agritourism offer, thanks to cooperation between individual 'tourist players'—gastronomy (the use of organic food), the hotel industry (Eco-hotels), or cross-marketing (the use of the Graubünden own brand on domestic products). In addition, most of the respondents agree as to the similarities between the tourism and agriculture industries, such as: seasonality, overlapping with the same branches of the economy and close links in the value chain. However, there are also concerns about cooperation between farmers and tourists (are tourists ready to pay higher prices for the price of recreation in a natural landscape and the opportunity to take advantage of a wider market offer and cooperate with local residents (Kuhnhenn and Simon 2013) (Table 1.7).

In summary, Graubünden has a very great potential for the development of tourism in the spirit of sustainable development, as evidenced by the above-mentioned example of synergy between organic farming and the tourism industry, which the region has not yet exhausted, and so far, uses it skillfully. Adequate emphasis on increasing cooperation between sectors and using expert support has a chance to create a great potential for expansion and market optimization. The region is required to have an appropriate individual strategy and appropriate approach to these issues, and its innovative activities may inspire other Alpine regions.

Table 1.7 Examples of the benefits of cooperation between the organic farming and tourism sectors

Advantages for organic farming	Advantages for the tourism industry	Advantages for both organic farming and the tourism industry
Achieving higher prices by selling groceries through tourist sales channels	Increasing competitiveness	Additional sales opportunities thanks to new marketing channels
Increasing business knowledge	Improving the image as an ecological industry	Acquiring new customer groups
Professionalization of marketing	Diversification of the range of services and products	Use of common strengths and the opportunity to specialize
Development of an additional second line of business with potential income and profits	Achieving higher sales by offering regional high-quality products	Reduction of the necessary workload thanks to cooperation and transfer of know-how
Increase in added value through on-site sales and food processing	Extending the stay of guests in the region	Swap competition with negative effects into mutual integration with positive effects

Source Own elaboration on basis: Kuhnhenn (2013)

Julian Alps region (Slovenia)—an example of sustainable settlement systems

The area of Slovenia is predominantly covered by mountains. There are approximately 700 km of marked trails and 170 mountain huts in Slovenia. The country is also famous for its large thermal and mineral water resources—there are 15 spas and a large number of karst caves—over 7000. In the north of the country there is the Julian Alps range with its highest peak, Triglav, 2864 m above sea level. The most important tourist centers are the Bled and Bohinj at their foothills and the Kranjska Gora ski center (Kurek 2004) (Fig. 1.14).

Slovenia as a former socialist state, in the second half of the 1980s, had changes in the economic development in mountainous areas, characteristic of the countries of this system. In mountainous regions, industry began to develop after the end of World War II in 1945. This was related to the regional policy of the state, which wanted to better use the potential of these areas, creating new jobs, which in turn led to an influx of people and an increase in urbanization in villages located in mountain valleys. Traditional subsistence agriculture has been preserved mainly in peripheral areas (Kurek 2004).

Currently, agritourism, which has been constantly developing for over 20 years, is becoming more and more important in tourism in Slovenia. This is the result of the policy of local authorities, which attach great importance to counseling, education and providing appropriate equipment for farmers to set up and run this type of farm. This policy is based on the following assumptions:

- treating agritourism as a complementary form of activity,

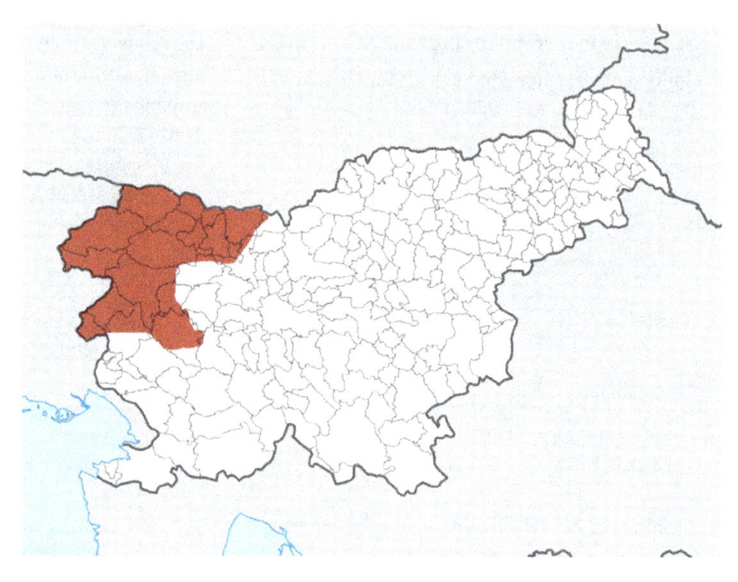

Fig. 1.14 Location of the Julian Alps (red color) on the map of Slovenia. *Source* www.commons.wikimedia.org/wiki/File:Slovania_location_map_-_Julian_Alps.png (09.06.2021)

- farms should be in a sufficiently good economic condition to undertake this type of activity,
- agricultural organizations are responsible for this type of farm, and training for farmers should be organized by agricultural advisory centers,
- appropriate financial and spatial policy should be carried out in order to maintain the proper architecture of buildings, infrastructure and landscape.

Barriers to the development of agritourism farms in Slovenia are the high costs of tourist investments in farms, high loan prices, and the poor level of education of the agricultural population (Kurek 2004).

It is also worth mentioning the phenomenon of second homes in the Julian Alps, where residents and owners of existing second homes are increasingly opposed to the construction of new buildings of this type, considering it to be harmful to the cultural landscape and the current traditional network of buildings. An example of preventing the proliferation of second homes is the Triglav National Park, where special legislation was passed in 1981 and 2010 to ban the construction of new second homes, but this did not completely eliminate the problem, but only reduce it. Many municipalities within the Park, such as Bohinj, Kranjska Gora or Bovec, have a problem with coping with the large, uncontrolled increase in the stock of second homes. Due to the problems with determining the correct number of them, it is difficult to assess the progress in reducing the phenomenon. Attempts to completely eliminate second homes can have negative consequences, so it is better to develop your holiday home stock in a sustainable manner with careful planning by your local authority. Their neglect of this issue may cause permanent ecological, environmental and cultural damage in this most popular mountainous region of Slovenia (Koderman and Pavlič 2019).

In Slovenia, while developing a strategy for the development of tourism and rural areas, it was noticed that in the face of ubiquitous progress, people are more and more willing to look for unknown experiences that can meet their increasing demands, and due to omnipresent globalization, there is an increasing interest in a simple form of recreation, returning to the roots, experiencing something 'first hand'. Therefore, recreation in the bosom of nature, in a rural landscape is becoming more and more popular (Mrak et al. 2012).

The main goals in the strategic documents for the development of Alpine rural areas have become:

- stimulating the protection of traditional types of settlements while adding new values to them,
- linking by means of rural development with cities with central functions,
- Attempting to specialize in agricultural activity and making it a guardian of a cultural region (Mrak et al. 2012).

When studying population dispersion in the Julian Alps, two types of settlement structures can be distinguished:

- rural tourist areas—located most often in hilly areas, have a well-developed tourist infrastructure, there is a phenomenon of emigration of peasant youth, which causes disruptions in the existing agricultural production and traditional settlement structure, they are often affected by problems related to the excessive increase in tourist opportunities, their main asset tourism is the existing architecture and high-quality surroundings, this type of area also includes central settlements (urban structures), which have changed as a result of urbanization pressure in the area of these structures,
- remote rural areas—located in some hilly and mountainous areas where forests and agricultural areas predominate with rare, highly scattered settlement, these are often areas with a significantly declining population, in extreme cases even completely abandoned, thanks to which they become the goal of developing tourism focused on the sustainable development of rural areas. It is the latter type of areas with scattered settlement that has become the main object of interest from the point of view of the development of sustainable, modern forms of tourism.

The Slovenian authorities proposed that the development of rural development should be guided by the principle of fusing old buildings with modern ideas, manifested by combining old buildings in the countryside—solid and durable buildings, but at the same time comfortable and safe, usually located around a courtyard, with a new type of high-quality connections with an identity causing mosaics from settlement structures, inhibiting wild growth, and imparting correlation between new spatial forms and the development of the cultural landscape. As a result of this mixture, self-organizing systems were created, where the most important thing in them is to be interesting and dynamic, while features such as stability or a specific term have been pushed to the background (Mrak et al. 2012).

An example of such a modern mountain tourist settlement with traditional settlement patterns is the part of the village of Livek called Nebesa, which combines the idea of scattered settlement with tourist content. Built at the intersection of roads leading to Venetian Slovenia on the Mediterranean Sea and Brda Goriška lying in the Alps, it is an ideal place for a developed complex with accommodation, for which the old ski infrastructure is used. The complex refers to the historical concept of scattered settlement, returns to the roots of the architectural heritage, while introducing modern architecture without calculating historical elements. The location in an open, well-developed space creates a unique rhythm of life in the complex, where each tourist can individually use the local, traditional system of connections between the six accommodation facilities and the surrounding farms, which are hiking trails and bicycle routes in a historic setting (Mrak et al. 2012).

In the region, the maintenance of balance in the cultural landscape is promoted, an example of which can be the concept and method of building mountain bivouacs. In the past, used by shepherds as a shelter for grazing animals in the mountains, these types of structures became a model for the functionality of modern architectural solutions in the design of settlements, being a sign of the former social life of the indigenous people. The campsite solution is considered spatially more democratic and ecological in comparison to the solutions from the twentieth century, which perfectly fits the tourist assumptions of the new millennium. As a result, the heritage of mountain bivouacs is under special protection due to their cultural value, and measures are taken to protect them against damage caused by uncontrolled tourist activity. Appropriate transformation of old, abandoned camps into modern tourist accommodation and shelters used by mountaineers and skiers is an opportunity to create an ecological accommodation base close to nature, which has a chance to be very popular among tourists. An example of such a camp is the Kotovo sedlo located under the Javolec peak, which is a popular destination for both summer and winter trips (Mrak et al. 2012).

In summary, the diverse patterns of scattered settlement are an important factor for the sustainable development of local space. They create opportunities for hard-to-reach areas to be used for tourist purposes, thanks to the rejection of the outdated model of mass tourism in favor of contemporary tendencies toward recreation focused on the proximity of nature, exploring the architectural and cultural heritage. This offers a chance for neglected, abandoned regions full of scattered residential centers located far from the main urban centers to shine and try to fuel the local economy with a developed tourism industry (Mrak et al. 2012).

1.7 Summary and Conclusions

The aim of this chapter was to introduce the concept of sustainable tourism and its indicators, to examine issues related to the development and condition of sustainable tourism in the area of the Alpine mountain range on the basis of examples of activities taken from the region and to answer two theses that the development of tourism in the regions selected in the study takes place in accordance with the principles of sustainable development and examples of activities undertaken in the tourism space of these regions are consistent with the principles of sustainable development. These goals were achieved in the theoretical and empirical parts of the work. The theoretical part explains the issues of sustainable tourism, which, although relatively new, is well described in the Polish literature on the subject, and there are many books and scientific articles available on this subject. The issue has many levels on which it can be considered, and there are also many related concepts with a similar meaning, which were also explained in the theoretical part. In the theoretical part, the literature on the subject was used, statistical data from official, verified sources were analyzed, and strategic documents at the international and national level were analyzed.

In the empirical part, examples of activities from the described region that are consistent with the principles of sustainable tourism development were examined. These activities cover various planes (spatial, economic, political), and each of the examples raises different issues related to the broad meaning of the concept of sustainable tourism. In the empirical part, the literature on the subject and the analysis of national and local strategic documents were used. After theoretical and empirical considerations, it was possible to answer the following questions about whether the development of tourism in selected regions is in line with the principles of sustainable development and whether the examples of activities undertaken in tourism space discussed in the paper are consistent with the principles of sustainable development. Both questions were answered in the affirmative, so the tourist traffic in the analyzed areas was carried out in accordance with the idea of sustainable tourism development.

Habitat loss is a huge problem in the modern world. As habitats decline, many species become endangered. In addition, theories about the onset of the COVID-19 pandemic indicate the reason for people entering natural habitats, where there live bats. Once upon a time these habitats were inaccessible to humans. Due to the demographic pressure, people began to expand their range and destroy the habitats of bats. The crossing of the species barrier for the COVID-19 virus caused the first epidemic wave in the city of Wuhan at the end of 2019, marking the start of the COVID-19 pandemic in 2020. Initial studies of the SARS CoV-2 coronavirus genome indicated that it came from bats from Rhinolophus affinis species living in the province of Yunnan, approx. 1000 km from Wuhan. A more probable theory is that it could have been passed on to humans through an intermediate host, i.e., a pangolin, which is the most trafficked mammal in the world (www.pzuzdrowie.pl).

Development of tourism was to bring a huge advantages to the developing countries. Meanwhile we can show a list of negative and harmful disadvantages such as: loss of biodiversity, littering by tourists, low-status jobs, creating a tourist economic monoculture or inflation pressure (Podhorodecka, Dudek 2019). However, the solutions, which are taken in well developed countries can bring more economical advantage. The COVID-19 pandemic show enormous changes in tourism movement. Not only the huge decrease in number of tourist arrivals but also the change in traveling (far travels are switch to smaller distance travels, foreign travels are change into domestic travels, long stays are change to short term stays). And of course the means of transport is also changing because people are reducing the travels by train and by planes and they are changing into private cars. During the COVID-19 pandemic times people prefer stay closer with family and they prefer isolated accommodation facilities (separate houses in environmental and isolated areas). That means that distance tourist destinations in developing countries are losing lot of incomes from tourism sector, because people afraid to travel abroad during uncertain times. Of course, the sustainable development of tourism can be remedium both for COVID-19 problems—overcrowded destinations and exploitation of environment. The concept of sustainable tourism and its indicators were shown with the examples of the Alpine Mountain range on the basis of examples of activities. The examples showed the principles of sustainable tourism development, which cover various planes (spatial, economic, political), and each of the examples raises different issues related to the broad meaning of the concept of sustainable tourism—case studies of High Alps Region (France), Graubünden region (Switzerland) and Julian Alps region (Slovenia).

References

Ashley S et al (2014) Morbidity and mortality of invertebrates, amphibians, reptiles, and mammals at a major exotic companion animal wholesaler. J Appl Anim Welfare Sci 17(4):308–321

Barnwell G (2003) Difficulties in paradise: the feasibility of sustainable development, New York, pp 3. www.caribvoice.org/Travel&Tourism/paradise.html

Bhuiyana MA, Rashid Khanb HU, Zamanc K, Hishand SS (2018) Measuring the impact of global tropospheric ozone, carbon dioxide and sulfur dioxide concentrations on biodiversity loss. Environ Res 160:398–411

Bourdeau P (2009) *Interroger l'innovation dans les Alpes à l'échelle locale*, "Semaine alpine 2008: innover (dans) les Alpes, France

Bromby M (2021) The Cayman Islands: paradoxes of insularity in the caribbean and other British Overseas territories. Liverpool Law Rev 42:35–49. https://doi.org/10.1007/s10991-020-09261-0

Bugdol M, Puciato D (2023) *Sposoby zapobiegania marnotrawstwa żywności w hotelach*, "Turystyka – zarządzanie, administracja, prawo", No 1/2023, Ministerstwo Sportu i Turystyki, Warszawa, s. 29–35. https://doi.org/10.61016/TZAP-2956-8048-3

Čech V, Javorská M (2023) Analysis of the Current State of Development in the area of Secured Routes (via ferratas) in Slovakia. Geogr Rev 19(2). ISSN 2585–8955

Cywiński P (2015) *Tourist Neo-colonialism as an Indication of the Future of Islands. The Example of Borobodur (Central Java).* Miscellanea Geographica. Reg Stud Developm 19(2):21–24

Cywiński P (2019) *Neokolonializm turystyczny jako zjawisko geograficzne i społeczne na wybranych przykładach z Indonezji*, rozprawa doktorska, Wydział Geografii i Studiów Regionalnych, Uniwersytet Warszawski

De la Vina L, Ford J (1998) Economic impact of proposed Cruiseship business. Annals of Tourism Res. 205–208; Pergamon

Duncan P, Graham P (2015) In: Darwin's sciences, Wiley & Sons, ISBN 978-1-4443-3035-9

Dwyer L (1998) Economic significance of cruise tourism. Annals of Tourism Res 25(2):393–415; Pergamon

European Commission (2017) Flash Eurobarometr 457. Report. Businesses' attitudes towards corruption in the EU. December, 'TNS Political & Social', European Commission

Gołembski G (2009) *Kompendium wiedzy o turystyce*, Wydawnictwo Naukowe PWN, Warszawa

Górniak J, Keller K (2008) *Rola systemów wskaźników w ewaluacji*, "Teoria i praktyka ewaluacji interwencji publicznych", Wydawnictwa Akademickie i Profesjonalne Spółka z.o.o., Warszawa
Haddad MN et al. (2015), Habitat fragmentation and its lasting impact on Earth's ecosystems. Sci Adv 1(2)
Impact Assessment of The COVID-19 Outbreak on International Tourism (2020) Unwto.org. UNWTO. www.unwto.org/impact-assessment-of-the-covid-19-outbreak-on-international-tourism
Jasiński M (2006) Aspekty monokulturowe gospodarki turystycznej. Zeszyty Naukowe SGH, Warszawa, Szkoła Główna Handlowa, Warszawa 20:92–106
Jędrusik M (2010) Geografia turystyczna świata Nowe trendy. Regiony turystyczne, Wydawnictwa Uniwersytetu Warszawskiego, Warszawa
Jędrusik M (2018) *Nosy Be (Madagascar) and the neighbouring islands versus tourism development*. Miscellanea Geographica—Regional Studies on Development 23(1): 1–10. ISSN: 2084-6118. https://doi.org/10.2478/mgrsd-2019-0001
Klimek K (2010) *Niekorzystne aspekty rozwoju turystyki w Szwajcarii*, 'Zeszyty Naukowe nr 825 Uniwersytetu Ekonomicznego w Krakowie'", Kraków, pp 65–83
Koderman M, Pavlič A (2019) *Vikendice u slovenskim Alpama s posebnim naglaskom na općinu Bovec*, Hrvatski Geografski Glasnik, pp 61–81
Konopinska N (2017) *Uwarunkowania zrównoważonego rozwoju turystyki na sudeckim przygranicznym obszarze górskim*, „Prace Naukowe Uniwersytetu Ekonomicznego we Wrocławiu", Wrocław, pp 146–156
Kowalczyk A, Derek M (2010) *Zagospodarowanie turystyczne*, Wydawnictwo Naukowe PWN, Warszawa
Kruczek Z (ed.) (2011) *Piloci i przewodnicy na styku kultur*, Wydawnictwo Proksenia, Kraków
Kuhnhenn U, Simon S (2013) *Biolandwirtschaft und Tourismuswirtschaft in Graubünden—Marktgerechte Nutzung von nachhaltigen Synergiepotenzialen*, Tagungsband der 12. Wissenschaftstagung Ökologischer Landbau, Bonn
Kurek W (2001) *Tendencje rozwoju turystyki w Alpach*, „Człowiek i przestrzeń: profesorowi Adamowi Jelonkowi w 70. rocznicę urodzin", Instytut Geografii i Gospodarki Przestrzennej UJ, Kraków, pp 185–193
Kurek W (2004) *Turystyka na obszarach górskich Europy wybrane zagadnienia*, Wydawnictwa Instytutu Geografii i Gospodarki Przestrzennej UJ, Kraków
Kurek W (2012) Regiony turystyczne świata. Część 1, Wydawnictwo Naukowe PWN, Warszawa
Milliken T (2014) Illegal trade in Ivory and Rhino Horn: an assessment report to improve law enforcement under the wildlife TRAPS project, USAID, Traffic
Morelli TL, Smith AB, Mancini AN et al (2020) The fate of Madagascar's rainforest habitat. Nat Clim Chang 10:89–96. https://doi.org/10.1038/s41558-019-0647-x
Moreno-Gené, J, Sánchez-Pulido, L, Cristobal-Fransi, E, Daries N (2018) The economic sustainability of snow tourism: the case of Ski Resorts in Austria, France, and Italy, "Sustainability", Lleida
Mrak G, Zavodnik Lamovšek A, Fikfak A (2012) *Trajnostni prostorski razvoj podeželja – poselitveni vzorci v Slovenskih Alpah, Arhitektura raziskave,* Architecture Research, Univerza v Ljubljan, Fakulteta za arhitekturo, Lublana, pp 10–21
Myga-Piątek U (2011) Koncepcja zrównoważonego rozwoju w turystyce, *Problemy ekorozwoju—problems of sustainable development,* Sosnowiec, pp 145–154
Niezgoda A, Jantczak K, Patelak K (2020) *Ekspertyza na temat wytycznych w zakresie zrównoważonej turystyki*, Ministerstwo Rozwoju Pracy i Technologii, Łodź
Number of bed places per km^2 in rural areas (2017) European Environment Agency. https://www.eea.europa.eu/data-and-maps/figures/number-of-bed-places-per-2/number-of-bed-places-per (05 May 2021)
Parks for Life: action for protected areas in Europe (1994) Gland, IUCN
Podhorodecka K, Dudek A (2019) Disadvantages connected with the development of tourism in the contemporary world and the concept of sustainable tourism. Problemy Ekorozwoju 14(2):45–55

Pullin AS (2004) Biologiczne podstawy ochrony przyrody, Wydawnictwo Naukowe PWN, Warszawa
Shakeela A, Weaver D (2018) *Managed evils of hedonistic tourism in the Maldives: Islamic social representations and their mediation of local social exchange*, Annals of Tourism Res. 71:13–24
Trykowski P, Podhorodecka K (2019) *Postawy Polaków wobec norm społecznych w Zjednoczonych Emiratach Arabskich i ich wpływ na turystykę*, 'Turystyka kulturowa', nr 2/2019, ISSN: 1689-4642
Yearbook of Tourism Statistics Data 2015–2019 (2021) UNWTO, Madrid
Zaręba D (2000) *Ekoturystyka. Wyzwania i nadzieje*, Warszawa, Polskie Wydawnictwo Naukowe

Chapter 2
Selected Social Issues in the Contemporary World—Global Determinants and Trends in International Migration

Barbara Dominika Jaczewska

Abstract Human mobility is currently one of the central topics of interest for researchers around the world. This part of the book describes challenges connected with international migration, and the main goal is to present forces shaping the contemporary migration movement, data on the scale, changes in the number of people, age and sex structure of migrants participating in international migrations, and preliminary findings of the influence of the COVID-19. The second part of the chapter concerns the history of migration to Europe as an example of the region that transformed from predominantly emigration into immigration within the last 60 years. It describes also the so-called 'European migration crisis' of 2015 and the crisis triggered by Russia's aggression against Ukraine in 2022 as unexpected events shaping migration features. Focusing on selected issues author's intention is to counter stereotypical views on contemporary international migration. In many countries, we can see an increase in citizens' concern about the influx of migrants, which leads to a radicalization of discourse and attitudes, but often results from ignorance of the history of migration, its actual scale and nature. The feeling that migration is increasing is often associated with a strong 'visibility' of some migrants, who are the subject of media interest of politicians, as well as the bargaining chip of election campaigns and ideological projects. The chapter presents data and explanations for some controversies so that everyone can form their own opinion.

2.1 Introduction

Human mobility is currently one of the central topics of interest for researchers around the world. It takes on different forms, including daily, circular mobility, migration, and tourism. The relationships and overlaps presupposed by these flows make them

B. D. Jaczewska (✉)
Faculty of Geography and Regional Studies, University of Warsaw, Krakowskie Przedmieście 30, 00-927 Warsaw, Poland
e-mail: bgibki@uw.edu.pl

K. Podhorodecka and T. Wites (eds.), *Global Challenges*, Springer Geography,
https://doi.org/10.1007/978-3-031-60238-2_2

tough to distinguish, for example, tourism is a form of mobility of variable duration, which in turn generates other forms of migration, such as those initiated by the demand for services for tourists, those linked to second houses, those dependent on seasonal cycles in the labour market, those connected to changes in lifestyle, and those related to the new habits of elderly populations and pensioners. On the other hand, many forms of migration generate flows of tourists, as immigrant communities can become hubs for such flows while simultaneously activating return tourism streams for visiting friends and relatives, and for maintaining a relationship with the country and culture of origin. These overlaps are even more visible when examining the migration of various groups, like highly skilled workers or ex-pats, or those resulting from a change in lifestyle and adopting a mobile lifestyle. People do not necessarily have to be centred in one place, they can also have translocal or multilocal life and alternate between, or simultaneously realize their intentions in, two or more places (Jaczewska 2023). Current studies also emphasize a systematic neglect of the causes and consequences of immobility that hinders attempts to explain why, when, and how people migrate or not and despite often used statistics that show that one in seven people are on the move, not so often researchers ask the question why, in our age of migration, six out of seven are not. The mobility and immobility nexus can be studied from two perspectives: as a result of structural constraints on the capability to move and/or as a reflection of the aspiration to stay. In this sense, a variety of potential explanations for the aspiration to stay, highlight the often overlooked 'retain' and 'repel' factors and economic 'irrationality' that also shape migration decision-making, particularly aspirations related to family or community that may vary by gender or social group.

All these examples emphasize that current mobility and migration processes are extremely diverse, multidimensional, and far from unambiguous. We are seeing both a continuation of established processes, which are taking on new forms and previously unknown types of population movement resulting from new economic, cultural, and environmental changes, as well as military conflicts. Demographic changes, globalization, and environmental degradation mean that migration pressures across borders will likely increase in the coming decades and migration will continue to be one of the most important issues of the global policy agenda as it generates enormous economic, social, and cultural implications in both sending and receiving countries.

This part of the book describes challenges connected with international migration, and the main goal is to present forces shaping the contemporary migration movement, data on the scale, changes in the number of people, age and sex structure of migrants participating in international migrations and preliminary findings of the influence of the COVID-19. The second part of the chapter concerns the history of migration to Europe as an example of the region that transformed from predominantly emigration into immigration within the last 60 years. It describes also the so-called European migration crisis of 2015 and the crisis triggered by Russia's aggression against Ukraine in 2022 as unexpected events shaping migration features. I concentrate my consideration on a wide and diverse group of migrants, which is defined as '*any person who changes his or her country of usual residence*' (UN DESA 1998). Stocks were defined as 'the total number of international migrants present in a given country at a particular point in time' (UN SD 2017). The selected characteristic features of labour migrants (highly skilled and low-skilled), refugees and asylum seekers, family members, and return migrants, both permanent and temporary are emphasized. Internal migration is not included in the analysis, but that does not imply that it is in any way less important than international movements.

This part of the book by no means exhausts the topic of international migration. Focusing on selected issues Author's intention is to counter stereotypical views on contemporary international migration and the migration situation in Europe. In many countries, we can see an increase in citizens' concern about the influx of migrants, which leads to a radicalization of discourse and attitudes, but often results from ignorance of the history of migration, its actual scale and nature. The feeling that migration is increasing is often associated with a strong 'visibility' of some migrants, who are the subject of media interest of politicians, as well as the bargaining chip of election campaigns and ideological projects. The chapter presents data and explanations for some controversies so that everyone can form their own opinion.

2.2 The Forces Shaping International Migration

This subsection reviews a range of drivers or factors[1] that encourage mobility and that shape the forms of international movements. We can look for the reasons for the increase in mobility in many processes, but only a few of them will be clarified in this section. I point out five aspects, which are connected with economic, social, political, technological, and environmental changes. All these factors create interconnected conditions, which are frames for the complex, interrelated, variable, and contextual development of international migration movements, which is why the description will be inevitably rather brief and superficial.

Researchers emphasize, that one of the most important changes to affect the increase in migration is **economic transformation**, which is related to global economic integration. Advocates of neoliberal globalization (especially since the mid-1970s to 2007) have persuaded that it would lead to faster economic growth in poor countries, and in the long run, to poverty reduction and convergence with richer countries. The opposite was the case (Milanovic 2007). Castles (2013) indicates that the claim of reducing inequality was 'a main element of political legitimation because it underpinned the principles of 'open borders' and level playing field'. The author specifies moreover, that flows across borders of commodities, capital, technology and labour were meant to secure optimal allocation of resources and to ensure that production factors could be obtained at the lowest possible cost. The continually changing poles of economic activity have brought about, in their wake,

[1] The term driver or factor do not imply determinism. Except in the case of emergency migration to escape violence or disaster, there is always an element of choice or agency in the decision to migrate and the majority of people chose not to migrate. The drivers are perceived as elements that increase the likelihood that people will decide to migration in search of a better life.

dynamic flows of people to new centres of development, but also the collapse of traditional economic systems (agriculture, crafts, and trade), which, in turn, forced further migrations (Jaczewska et al. 2018). Neoliberal globalization has caused migratory movements to affect a greater and greater number of states simultaneously, and more strongly. Global phenomena and local development became strongly interrelated, with human mobility being one of the most significant expressions of this interrelationship. Liberalization of flows was never complete and the hypocrisy is greatest concerning flows of people, where control of movements across borders is often seen as an important part of nation-state sovereignty. International migration nowadays is driven mostly by labour market imbalance, wealth inequality, and economic instability in countries of origin.

As we have more and more global economic connections, migration has also become a branch of worldwide, international business. De Haas et al. (2019) point to the existence, and also increasing numbers, of various groups who earn a living from organizing migration flows, which they have dubbed 'the migration industry'. Among these groups are travel agents, employee recruiters, intermediaries, translators, real estate agents, and lawyers, but also smugglers and document counterfeiters. The development of the migration industry is an inseparable aspect of social networks and transnational connections, but it is also a driving force for migration streams and can be used to explain why migration can become self-perpetuating and occurs even if economic, political, and social conditions change. The authors emphasize that migration agents form an international network that supports the global labour market, and that over time, the migration industry could become a major driver of migratory flows. The existence of a strong migration industry is used to explain the failures of official migration policies. Government policy aimed at curtailing the flow of migrants comes into conflict with the economic interests of migration agents, who work to continue it, even if the form of migration changes, for example, legal recruitment will be replaced by illegal recruitment.

Migration is not an isolated process, and almost always population flows are prompted by **social transformation**. Modifications in migration are closely related to social transformation and social transformation's influence on migration, so an increase in the scale of migration can be both a cause and an outcome of social transformation. The notion of social transformation signifies the profound structural modification of societal relations. It directs our attention to the structural embedding of societal shifts, to globalized relations of power and inequality, and the complex interplay between social, cultural, political, and economic dynamics. According to this understanding, contemporary changes in migration are embedded in social transformation processes, which are regarded as processes that are continuously produced and generated. Social transformation processes are necessarily located in time and space. Therefore, special attention should be given to the changes themselves, but also to the configurations in which societal changes unfold (Amelina et al. 2016). Such an approach emphasizes that, while trying to understand the complexity, interconnectedness, variability, and context of migration, we have to be sensitive to both the strategic agency of social actors and the structural frameworks they are confronted with. Quite often migration policies and public perceptions concentrate on economic aspects of migration and ignore that migrants are members of families and communities. Migration in that sense is a social process, in which the participants undergo processes of change, and in turn, act to change (transform) the conditions and practices that they encounter. Massey, in his very well know concept of the cumulative causation of migration, indicated the importance of social transformation. He stated that 'every act of migration makes a change in the social environment, within which migration decisions are taken further, usually increasing the likelihood of subsequent flows' (Massey et al. 1994). These social dynamics play an important role in shaping the volume and forms of international migration. Newcomers who arrive at new destinations from many countries of birth that differ significantly in economic, social, and cultural terms, contribute to the increase in the importance of a variety of migration networks based on family and ethnic ties, as well as professional connections. Migration policies often ignore the social nature of the migratory process and concentrate on the allocation of different individuals into specific administrative categories, which may not correspond to economic and social realities and result in irregular migration or experience of exploitation and insecurity that can lead to resistance. We are now witnessing an upsurge in protest movements of disadvantaged and vulnerable groups such as migrant women, irregular workers, and ethnic and racial minorities. Resistance to the structural inequality in incomes and human security inherent in the global labour market makes it clear that migrants are not passive victims, but are taking an active part in social transformation and can develop new forms of social and political actions.

Political changes, the securitization and the politicization of migration are also not without significance in shaping migration movements. Political discourses play a key role in migration processes: they signify the dominant narratives about society and its social reproduction, they define the understanding of major problems, and they are the basis on which shared political objectives are formulated. Politicians in labour-importing countries, aware of popular suspicion of immigration respond with rhetoric of national interests and control. They try to resolve the inconsistency between strong labour demand and public hostility to migration by forming entry schemes that encourage the legal entry of highly skilled workers, while either excluding lower-skilled workers or regulating them through temporary employment systems. Since the labour market demand for the lower skilled is strong, millions of migrants are pushed into irregularity (Castles 2013). Governments often do not perceive it as a problem in times of economic growth, and then tighten up border security and deport irregulars in times of recession. Politics also influences the increase of the securitization of political actions on migration and ethnic minorities. For example, the restrictive tendencies in current European migration regulations are embedded in securitization discourses: the political definition of migration (in particular, for non-EU countries) as an existential threat to social security and public order is a necessary condition of the remarkable change that manifests itself in the manifold deprivation of migrant rights in Europe (Huysmans 2006). This approach ignores the reality that migration and refugee flows are often the results of the fundamental lack of human security in many poorer countries. Such insecurity (poverty, hunger, violence and lack of human rights) is not a 'natural' condition but is a result of past practices of colonization and more recent economic and political power structures, which have exacerbated inequality in many places (Castles 2013).

Nowadays, the term politicization of migration is used to describe a situation in which a part of politicians and the public are convinced that migration is critically important for the development of the international community. On the one hand, we can see that in recent decades it is no longer possible to control migration only by utilizing the internal legislation of individual countries and migration governance has now become subject to numerous regulatory measures at the transnational scale, including by the United Nations and European Union. On the other hand, national policies, bilateral agreements, and regional relations are increasingly being modified under the influence of migration (i.e. Turkey–EU relations). Of course, the influence on migration by those policies, and their effectiveness, is the subject of numerous considerations (Czaika and de Haas 2013).

Politics, the securitization or the politicization of migration influence the way migration is perceived. Proposed immigration policy can create legal frameworks for migration, remove barriers to migration for certain categories of migrants, or implement very restrictive policies. Integration policies can facilitate the functioning of migrants in society. Migration policies can increase human insecurity, especially when states refuse to create legal migration systems despite strong employer demand for workers. In such situations, migrants experience high levels of risk and exploitation. As a consequence, smuggling, trafficking, and lack of human and worker rights are the fate of millions of migrants.

Another significant factor is connected to the ongoing **technological changes** (related to the third and fourth industrial revolutions), which are especially visible in communication technologies, transport, and the development of mass media. New communication techniques and cheaper and faster transport have made travel accessible and easier to realize for a larger group of people. The ongoing automation of traditional manufacturing and industrial practices using modern smart technology has created changes in the labour market and increased the role of highly qualified specialists in the migration movement. What seems to be extremely important in the context of contemporary migrations is the growing role of mass media in influencing migration. The notion of mass media refers to a diverse array of media technologies that reach a large audience via mass communication. These technologies include broadcast media (i.e. film, television), digital media (Internet, mobile mass communication), outdoor media (i.e. advertising and billboards), and print media (i.e. newspapers). The mass media play a special role, being not only a source of up-to-date information but also an important tool for shaping the reality around us. They increasingly affect human spatial behaviours, choices, and attitudes. The mass media's impact on spatial perception is ambiguous; not only can the Internet, press, radio, and television provide the general public with most of its information about places and geography, thus shaping our perception of places, but also they create our image of attractive places to live, though not always consistent with the reality. Nowadays, a new package of interactive, participatory and networked spatial media is reshaping spatial knowledge and helping us while we are on a move. In 2015, while we were able to observe in news an increase in the number of refugees and asylum seekers trying to move to EU countries, some of the voices within the public debate indicated that those people who were arriving were not poor as they had smartphones. Access to smartphones has become extremely important and is one of the basic elements of travel as it provides access to information and allows people to respond to changes in the situation.

Media are often believed to be 'an additional factor' in shaping public attitudes and in producing narratives that construct the receiving society's image of immigrants (Caponio and Cappiali 2016). The mass media can help us understand migration movements and their consequences, or it can create a negative picture of migration and the migrant population. While media coverage differs from country to country and adapts to a positive or negative political environment towards migrants and migration, the role of the media remains key in shaping spatial behaviours, and public perceptions, and creating the conditions for political changes in migrant-related policies. The mass media have begun to influence migration decisions since they allow migrants to pay virtual visits to potential destinations, more easily make arrangements for travel, accommodation, and work, and then, when in the host country, maintain contact with their community and transnational networks. Mass media also influence how migrants are perceived, sometimes used to disinform, and spread lies (Neidhardt and Butcher 2022). The aspect that should be further studied is the ethical role of mass media in the case of international migration. Starting in 2015 the Ethical Journalism Network (EJN) tried to answer the question of how well migration is being reported in different parts of the world.[2] This contributed to the establishment of a five-point guide for migration reporting listed below:

1. Facts, not bias (use accurate, inclusive and fact-based reporting),
2. Know the law (understand and use terms on migrants in the right way, accordingly to the law),
3. Show humanity (avoid victimization, oversimplification, and the framing of coverage in a narrow humanitarian context that takes no account of the bigger picture),
4. Speak for all (include migrant voices, listen to the communities etc.),
5. Challenge hate (avoid extremism, be careful with inflammatory content) (Ethical Journalism Network 2017).

The last group of factors that is already influencing and will continue to influence migration is connected with **environmental changes**. Environmental migrants are people, who 'for compelling reasons of sudden or progressive change in the environment that adversely affects their lives or living conditions are obligated to leave their habitual homes, or choose to do so, either temporarily or permanently and who move either within a country or abroad' (Sobczak-Szelc and Fekih 2020; Serraglio et al. 2019). Within this meaning, migration is a long-standing form of environmental adaptation, and yet it is only one among many forms of adaptation. Environmental

[2] In 2017 EJN published study on how media on both sides of the Mediterranean report on migration for the International Centre for Migration Policy Development (ICMPD) and co-authoring a chapter to an IOM publication on improving data on missing migrants. In 2017 the EJN authored Media and Trafficking in Human Beings—Guidelines for the ICMPD.

changes can be expressed as all types of disturbances of the environment caused by human influences or natural ecological processes. Nowadays, climate change is seen as a critical, global push factor for migration flows, along with political conflict and economic inequality, and has been recognized as an emerging important driver of mobility by the Agenda for Humanity, the 2016 United Nations Summit for Refugees and Migrants, and the Global Compact for Migration and the Global Compact on Refugees (Stoler et al. 2021). We have to be aware that environmental changes can occur suddenly (so-called fast-onset changes) in the form of earthquakes, hurricanes, floods, and tsunamis; or they can occur progressively (so-called slow-onset changes), as in the case of gradual desertification or land degradation. In either circumstance, environmental changes may generate migration. The interconnections of environment and migration are usually better recognized for fast-onset changes. This thesis can be confirmed by the numerous publications regarding migration due to Hurricane Katrina in 2005, or the Coastal El Nino 2017 in Peru. While such events draw substantial media and scholarly attention, they fail to account for climate-related migration beyond the making of a disaster. Usually, progressive changes are more complex and it is very hard to describe precisely, the interactions between environment and migration or the mechanisms that cause migration, as they are the result of a variety of layered causes—economic, social, and political—which are stressed by changing environmental conditions as well as, frequently, by developmental and demographic conditions (Stoler et al. 2021). The patterns of movement of environmental migrants can also vary. Forced migration might result from an environmental catastrophe such as a flood or tsunami, or a government-instigated relocation, while a more gradual process of migration could be caused by slow-onset environmental deterioration. In the coming years, migration flows related to climate change are expected to increase, particularly in the world's poorest countries (Olsson et al. 2014; Cho 2019).

The distinguishing feature of our times is the global nature of migration, whose patterns are rooted in history and whose contemporary contours are affected by various social, economic, political, technological and environmental factors. It is impossible to predict what kind of forces will shape international migration in the future. Knowledge of current trends still does not help with speculation on unexpected events that can lead to major shifts. A good example is the pandemic COVID-19 which influenced international movement and almost stopped international mobility for some months.

2.3 The Scale of International Migration

A further part of the argument regarding contemporary global international migrations deals with the most important statistics and looks for an answer to the question of whether we live in an age of unprecedented migration, where migrants move and what are the countries of origin. The significance of contemporary migration is not due to the statistical size. For unlike humankind's earlier migration, today's people movements are concentrated in countries which are highly saturated in terms

Table 2.1 Key facts and figures from World Migration Reports 2000 and 2022

	2000	2022
Estimated number of international migrants	173 million	281 million
The estimated proportion of the world population who are migrants	2.8%	3.6%
A country with the highest proportion of international migrants	United Arab Emirates	United Arab Emirates
Number of migrant workers	–	169 million
Global international remittances (USD)	128 billion	702 billion
Number of refugees	14 million	26.4 million
Number of internally displaced persons	21 million	55 million

Source Based on McAuliffe, M. and A. Triandafyllidou World Migration Report 2022. International Organization for Migration (IOM)

of population, economy, and civilization. In addition, (and more importantly, as it seems), the significance of these migration flows arises from the complexity of the effects which potentially accompany them.

According to UN DESA (2021), the number of people living outside their country of birth or citizenship reached 281 million in 2020.[3] Since the 2000s the number of international migrants grew by 108 million (from 173 million in 2000). The number of international migrants has expanded faster than the global population, and the share of international migrants as part of the total population increased from 2.8% in 2000 to 3.2% in 2010, and further to 3.6% in 2020 (UNDESA 2021). The growing number of international migrations and the growing share of migrants as part of the total population since the 1990s can be used to confirm the thesis that the scale and increase in international migration are unprecedented; but on the other hand, the share of international migrants as part of the total population is still relatively small and has not gone up significantly (Table 2.1).

Some researchers point out that the absolute number of migrants has gone up because the world's population has increased, and that all forms of mobility have gone up, from travel, commuting, and business travel to tourism, but, migration, in terms of people changing residency, has been remarkably stable. With the growing interconnectivity between the world's societies, people's awareness of other cultures will only increase in the future, but this does not necessarily have to translate into an unprecedented increase in migration. Analysis of data by de Haas etal. (2019) questions the widespread idea that the volume, diversity, and geographical

[3] The Author is aware of the imperfection of the UN DESA statistics, still, they can be used to describe general trends. In reality, both the number of international migrants and their population growth on a global population scale are probably higher, as the data do not include undocumented migrants or migrants residing outside their country of origin but having temporary resident (or worker) status there. Similarly, in countries where statistics are based on citizenship, a migrant who has acquired citizenship is not counted in the statistics.

scope of international migration have increased significantly. They are convinced that global migration has thus not accelerated, but post-WWII migration shifts have been predominantly directional. Instead of asking: will there be many more or fewer migrants? A more relevant question would be to ask: where will future migrants come from and go to?

Of all migrants in the world, most people live in Europe (31%), and the least (not including Oceania) in Latin America (5%, although it has been growing dynamically in recent years). In the case of Latin America, the share of migrants is half the share of its population in the Earth's population. In the case of North America, the share of migrants (21%) is four times greater than the share of the population, while in Europe this share (31%) is three times greater. This will prove that these are the most attractive continents for people moving around. In 2020, just 20 countries hosted about two-thirds of all international migrants. The USA continued to be the most important destination country for international migrants with 51 million migrants in 2020, which is 18% of the world's total. In second place was Germany with 16 million international migrants, followed by Saudi Arabia (13 million), the Russian Federation (12 million), and the UK and Northern Ireland (9 million) (UNDESA 2021). Although the USA's number one spot has been stable for many years, the following ranks have changed since the 1990s. Western European countries, especially Germany, and Persian Gulf countries like Saudi Arabia, have strengthened their position, while the Russian Federation as a destination country has become less important. We can also observe changes in the position of countries of origin. Between 2000 and 2020, the size of the migrant population abroad grew for nearly all countries and areas of the world (UNDESA 2021). In 2020, India had the largest diaspora (the largest number of persons that live outside of their country of birth), which is estimated at 18 million (increasing by 10 million between 2000 and 2020). Other states with a large diaspora are Mexico and the Russian Federation (11 million each), China (10 million), and the Syrian Arab Republic (8 million). The position of India has increased, while Mexico and China have decreased. For the Syrian Arab Republic, the increase in their transnational population was primarily due to the large outflow of persons displaced across borders. Of the 20 countries or areas with the largest number of international migrants abroad in 2020, six were from Europe, five were from Central and Southern Asia, and four were from Eastern and Southeast Asia. The data indicate the growing importance of Asian countries as the main region for migrant origin (Figs. 2.1 and 2.2).

While most countries or areas saw the size of their transnational communities increase between 2000 and 2020, in 12% of all countries the size of their transnational populations shrunk. In some countries, including Angola and Serbia, this decline resulted from the voluntary return and repatriation of refugees to their home countries in post-conflict years. In others, such as Belarus and Georgia, the decline was primarily due to the older age structure of their transnational community and the mortality rates associated with this age structure (UNDESA 2021). For most countries, the size of their populations living abroad is quite small, relative to the native-born population still in their country of origin.

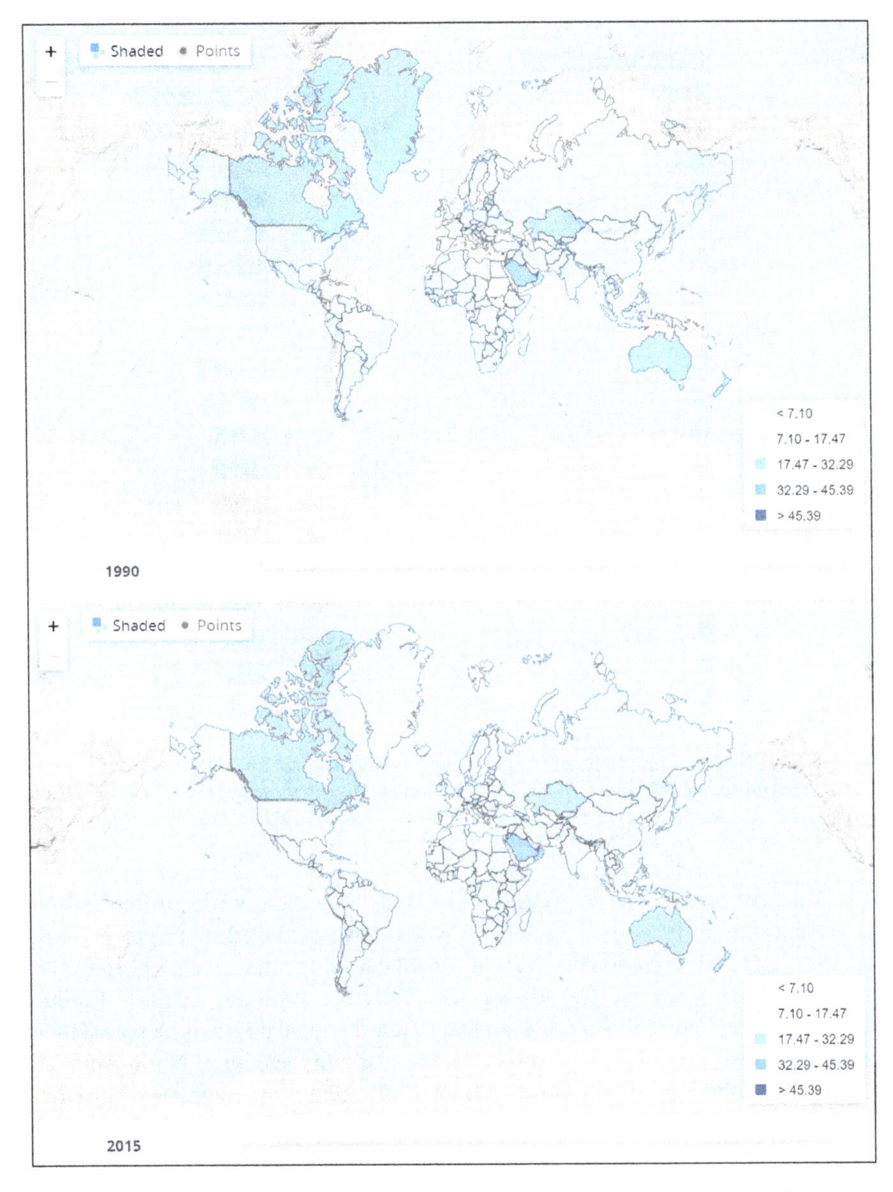

Fig. 2.1 International migrant stock (% of the population) in 1990 and 2015. *Source* World Bank, www.data.worldbank.org

Fig. 2.2 Top 20 destinations (left) and origins (right) of international migrants in 2020 (millions). *Source* McAuliffe and Triandafyllidou (2021). *World Migration Report 2022*. International Organizations for Migration (IOM), Geneva

It is almost certain that the future outlook for migration will be different from the current one. Fifty years ago, nobody would have predicted that Europe would become the global migration destination, which currently attracts more people than, for instance, the USA and Canada together. Between 2000 and 2020, in Europe, North Africa, and Western Asia, the number of international migrants increased most significantly (30 and 29 million, respectively), while in the case of North America, it was by around 18 million. About 9 million of the 29 million migrants received by North Africa and Western Asia were refugees or asylum seekers. If current trends continue, within the next decades North Africa and Western Asia are likely to surpass North America as the region with the second-largest number of migrants in the world (UNDESA 2021). This shift—hard to foresee twenty years ago—reflects the growing diversification of economic opportunities available to migrant workers. In the future, destination countries will be under greater competition while implementing migration systems that attract migrants, especially highly skilled migrants (Boeri et al. 2012; Czaika and Parsons 2017). Likewise, Europe, North Africa, and Western Asia, along with several other regions, have experienced an increase in the size of their migrant populations over the past two decades. This may also indicate a shift in sources of migrant labour. Areas such as sub-Saharan Africa, Eastern and South-East Asia, Latin America and the Caribbean, each received over 8 million migrants during that period. In Latin America and the Caribbean, the size of its migrant population more than doubled between 2000 and 2020. This was mostly, but not only, connected to an inflow of large numbers of displaced people from the Bolivarian Republic of Venezuela (WMR 2019).

While discussing the new pattern of migration, let us look at some examples, such as the migration of Africans to China. Many people think that most Africans who want to leave the continent all go, or all want to go, to Europe. First of all, the majority of Africans move to prosperous countries within the continent, and only some decide on intercontinental migration. What few people realize is that more and more Africans explore destinations outside of Europe, and China is one of those new destinations. The African diaspora in China emerged during the late 1990s economic boom, and thousands of African trades and business people, predominately from West and North Africa, arrived in China and created an African community in the middle of southern Chinese metropolises. Guangzhou is the best-known example. The number of people in the African community grew especially within the first decade of the twenty-first century but declined in 2014 due to strict immigration policy enforcement by Chinese authorities and economic pressure in home countries. It is assumed that this decline described only legal residents, and there are no equivalent figures for 'overstayers', so-called 'voluntary undocumented' migrants, who hid or lost their passports to become untraceable if caught. COVID-19 and actions taken by the government to trace the spreading of the virus make life difficult for migrants without documents. Without a legal residence, foreigners cannot apply for the Alipay Health Code, a system that assigns a colour code to users indicating their health status and determining their access to public spaces such as malls, subways, and airports. It may become impossible to live without documents, and African overstayers may be among the first to feel the consequences of this new norm in China. By 2020, there are an estimated 500,000 Africans living in China, with the majority residing in Guangzhou (13,652 in 2019).

Another example of changes is related to migration transformation countries. Over the last decade, countries, for example, Mexico and Turkey, have played the role of emigration countries and are important sources of both labour and family migration in North America and Europe. In the case of Mexico, according to the 2020 National Census, more than 1.3 million foreign-born people are living in Mexico, up from 900 thousand in 2010 and 400 thousand in 2000. Around 70% of foreigners living in Mexico come from neighbouring countries (the USA and Guatemala), while other important communities are from the Bolivarian Republic of Venezuela, Colombia, Honduras, Cuba, Salvador, and Argentina. The rest of the immigrants come from other non-Hispanic nations. The number of migrants and refugees residing in Turkey stands at 3.9 million. In 2019, refugees of the civil war in Syria had the highest registered number of refugees (3.6 million), but Turkey, even before the Syrian refugee crisis, had already become a destination for labour and other migrants, while the number of Turks leaving Turkey had gone down. One of the examples was the migration of retirement migrants, both lifestyle migrants or amenity-seeking migrants, and returning retired labour migrants who came as guest workers to the EU. Nowadays Mexico and Turkey are good examples of countries that experience migration transformation,[4] and who have become important countries for transit migration and countries of destination.

As mentioned earlier, migration does not affect spaces in the same way, and not every country experiences increases in migration. While in 179 countries we observed an increase in the number of migrants between 2000 and 2020, in 53 countries or areas, the number of international migrants declined. Among the countries that experienced declines, it is worth mentioning Armenia, India, Pakistan, Ukraine, and the United Republic of Tanzania.

To sum up: we live in an age of unique migration also because of the speed of the changes taking place and their complex nature. The changes in scale and pace of international migration are notoriously difficult to predict because it is closely connected to acute events (such as severe instability, economic crisis, or conflict) as well as long-term trends. We also know from long-term data that international migration is not uniform across the world but is shaped by economic, geographic, demographic, and other factors, resulting in a distinct migration pattern; so in the future, migration trends will continue to change creating new unpredictable constellations. For example, COVID-19 has reduced the number of international migrants and has influenced all forms of human mobility, including the international migration discussed in this chapter. Preliminary estimates by UNDESA (2021) suggested that by mid-2020 the pandemic may have reduced the growth of international migrants by around two million. According to McAuliffe and Triandafyllidou (2021) because of COVID-19 restrictions in 2020, there were 1.8 billion air passengers globally which was a major decline from 4.5 billion in 2019. It is worth mentioning that number of refugees, and internally displaced persons (IDPs) grew between 2019 and 2020. The real impact of the pandemic on international migration, however, will only be determined in a few years. We still do not know how long the pandemic will affect our lives and whether or not new variants of the virus will permanently change the direction of migration or the scale of migration.

2.4 Women, Children and Elderly People in Contemporary International Migration

The demographic characteristic of migrants usually is not widely discussed in the media, when greater importance is attached to the political aspects of migration and the legal or irregular status of migrants. More and more studies concentrate,

[4] The concept of migration transition derives from a hypothesis proposed by Wilbur Zelinsky (1971) that states that changes in social mobility occur in parallel with the phases of demographic and economic change (the development from traditional to highly developed societies). Demographic and economic development is accompanied by changes in the destinations and intensity of migration flows (King 2012; Skeldon 2012).

however, on gender and migration and children and elderly migration, and those subjects will be discussed in more detail in this section. It is worth emphasizing that the act of migration affects different family members in different ways, depending on their current role, and in the long run, the decisions of one generation affect its descendants.

An important characteristic of contemporary international migration (especially economic) is its feminization. However, this issue needs to be considered not only in terms of the growing number of women and girls participating in migration but also the increased interest in the subject among researchers. The data prove that women's participation in migration was consistently high in the previous decades of the twentieth century (Kofman and Raghuram 2022), and the paucity of knowledge about women and experiences resulted from, among others, from recognizing their mobility within the category of family migration (considered less important than employment migration). During the period 1960–2015, the number of migrant women doubled, but there were also increases in both the number of migrant men and the world population. The relative participation of women in 2020 compared to 1960 increased from 46.6 to 48%. Nowadays the number of female migrants slightly surpassed that of male migrants in the case of Europe, Northern America and Oceania. The most important factor influencing this is the higher life expectancy of women among long-term migrants and the increasing demand for female migrants in care-related sectors. The number of male migrants exceeded the female share in Northern Africa, Western Asia and sub-Saharan Africa. This imbalance is due to greater demand from male-dominated industries and the dominance of temporary work visas and labour contracts targeting male workers.

The current feminization of migration relates to an increase in awareness of the characteristics of female migration and investigation of the causes and consequences of migration gender balance, which varies over time (Donato and Gabaccia 2015; Grabowska-Lusińska and Jaźwińska-Matylska 2013). The increase in the number of women and girls in migration is associated with several factors, including sex differentials in survivorship and migration policies. One factor is that women, including migrant women, tend to live longer than men and that is why the ratio of female to male migrants is higher in regions that have past experiences in permitting or encouraging migration for permanent settlement or family reunification. Compared to other regions, Northern America and Europe had a higher proportion of female and older migrants among their migrant population, due to in part such policies. Another factor is connected with the increasing gender-specific demand for care-related work, which results from population ageing and changes in labour force participation as a consequence of social transformation (Barone and Mocetti 2011; Cortés and Tessada 2011; Farris 2015). This care-related work, which was previously performed by native-born women, often without pay, is being increasingly taken up by migrant women from lower-income countries. Older women in societies of origin often play an important role in this so-called 'global care chain', serving as care-givers to children 'left behind' by female migrant workers (Bryceson 2019; Dolbin MacNab and Yancura 2018).

Why migration of women is also important? Women can be seen as agents of change. According to Antman (2015), women transform social, cultural and political norms and promote positive social change across households and communities. Moreover, they contribute to the economic development of their countries of origin and destination. The creation of opportunities for female employment can contribute to promoting gender equality and empowerment for women and girls in both countries of origin and destination (Antman 2015; Ferrant and Tuccio 2015). Notwithstanding, migrant women remain among the most vulnerable members of society. There are still barriers that prevent women from full participation in social, political and economic life (Weichselbaumer 2016; Hennebry, Petrozziello 2019). They often have lower employment rates than native-born women or male migrants and are paid less than their male counterparts. The wage gap persists as migrant women are less frequently promoted, they are less often appointed to senior positions and they tend to be clustered in specific, lower-paid occupational groups (OECD 2020). This is strongly related to jobs most often done by migrant women in the destination countries in the care sector, which is one of the sectors offering the lowest wages. Women more often suffer because of skills underutilization as a consequence of a lack of gender equality. Women also face gendered risks of violence, abuse or exploitation, including human trafficking (EIGE 2020). Most of the estimated 225.000 victims of trafficking between 2003 and 2016 were females (UNODC 2018).

In countries of origin, the linkages between migration and gender can have some negative social and economic consequences. The emigration of family members, for instance, can slow down the formal labour force participation of non-migrant women, especially those living in rural areas (Asiedu and Chimbar 2020). The emigration of women and other family members can have an impact on the mental health and well-being of children who remain home. Also, older persons are directly affected by migrations (Adhikari et al. 2014). All of those aspects show the importance of implementing efficient and gender-responsive migration policies. Around half of the governments (with available data reported) indicated that have formal mechanisms to ensure that migration policy was sensitive to gender issues. However, such formal solutions were still missing in a large number of countries, including many where women and girls outnumber men and boys in the migrant population.

An interesting form of female migration is marriage migration, which will be discussed based on the Asia example. In Asia, migration for marriage is one of the few forms of permanent migration available in the region (most Asian migration is temporary). It has a long tradition as in other regions in the world, the so-called matchmakers, have often helped and are still helping to find the right mate. In the case of Asian women, this type of migration become an international migration phenomenon. The increase in the number of marriage migrants is connected with the development of transnational contacts and social transformations. Asian women migrated as brides of American soldiers from the 1940s onwards, first from Japan, then Korea and Vietnam. In the 1980s, there was the phenomenon of Asian 'mail-order brides' in Europe and Australia. Since the 1990s, farmers in rural Japan and Taiwan have been finding marriage partners abroad due to the exodus of local women to the cities in search of more attractive living conditions. Early in the twenty-first century, brides for men in India were recruited in Bangladesh, and Chinese farmers sought wives in Vietnam, Laos and Burma. The migration of women from diverse countries in Asia such as China, the Philippines and Vietnam for marriage to men from richer Asian countries such as Japan, South Korea, Singapore and Taiwan was the most dynamic form of permanent migration in East Asia (Yang and Chia-Wen Lu 2010). Diasporas may also be the driving force behind marriage migration. An example may be the phenomenon of second-generation migrants living in Europe bringing their spouses from their parents' country of origin, as is the case, for example, among Turks and Pakistanis in Belgium and Denmark (Van Pottelberge et al. 2021). Nowadays, marital migration is also facilitated by global tourism and the development of online dating sites.

As in other types of women migration, marriage migration can have positive and negative consequences. Women that choose an international marriage can improve their economic status back home, and attain greater social status. It can be an important step in gaining autonomy in decision-making. However, research has documented also negative impacts because women although they pursue upward mobility and productivity in both the sending and the receiving societies, are often vulnerable and face economic, cultural, and legal constraints. As a result, are often in a trap of being a 'victim' and 'agent', negotiating and strategizing between the constraints and opportunities they face (Huang and Yeoh, Lam 2008).

Most international migrants (73%) are of working age between 20 and 64 years. Children and adolescents are underrepresented among international migrants: globally, less than 15% of all international migrants were under 20 years of age, compared to 33% in the overall population. In the case of children, it is worth stating that they take part in migration as family members, but also alone as unaccompanied minors. The number of migrants who are unaccompanied minors has grown on an unprecedented scale in recent years. Of those seeking asylum during the migration crisis in the European Union in 2015, 88.300 were unaccompanied minors, while from 2008–2013, this number ranged between 11.000 and 13.000 annually. In 2021, 23,255 asylum applicants were considered to be unaccompanied minors, up 72% compared with 2020 (13,550). To a large extent, this increase was caused by the rise in the number of unaccompanied minors from Afghanistan (12,270 in 2021 compared to 5495 in 2020) (Eurostat 2022). The number of unaccompanied children crossing the U.S.-Mexico border increased since 2009, drawing the attention of the U.S. government, media, and public. In the USA in 2014, U.S. Customs and Border Protection apprehended more than 66.000 unaccompanied children at the border with Mexico (up to 2012 this number did not exceed 20.000 annually). Among those apprehended, most came from Central America (mainly El Salvador, Guatemala and Honduras) and the second important source was Mexico. After the implementation of more restrictive rules, the number dropped until 2019 when we could observe once again a significant increase. UNHCR (2022a, b) estimates, that the number of forcibly displaced children reached an estimated 36.5 million in 2021, meaning that children constituted approximately 41% of forcibly displaced people.

Children constitute a uniquely vulnerable migrant population. They travel alone to join family members who are abroad, to escape persecution in their home country, or as victims of human trafficking, smuggling or gang violence. Although there is consensus that children travelling alone should be treated in a manner appropriate to their age, non-always governments can ensure protection and offer residence opportunities. Researchers indicated a few reasons why this type of migration is so specific. The first is that we consider people who need care. Migrant children cannot simply get jobs and fend for themselves upon arrival in a new country. They need to be housed, educated and fed. This drives to the second aspect that this care is costly and complicated, especially as we have migrant children who experience violence, a trauma in their own countries. In light of modern trends in migration, we can assume that there will be continuing growth in the number of unaccompanied migrant children, and that helping and integrating them will be an important challenge for host countries.

What is quite interesting there is an overrepresentation of older migrant groups among international migrants compared to the total population. In 2020 12% of international migrants worldwide were at least 65 years old, compared to 9% of the total population. The difference was more pronounced among females and is probably due to a combination of already discussed gender factors. The age distribution of immigrants upon arrival, which tends to be concentrated between the ages of 20 and 40, is one of the main reasons. Moreover, children born to international migrants in countries of destination are not ‘foreign-born’ and are therefore not classified as migrants in statistics which also contributes to the distinctive age pattern of international migrants compared to the overall population. Another aspect is connected with the world of ageing. The world’s population aged 65 years or above is projected to increase from 700 million in 2020 to 1.5 billion in 2050, while the median age of the world population is projected to increase from nearly 31 years to over 36 years, so probably the number of older international migrants because of ageing of the world population will also increase.

In the context of the diversity of migration and the varying motives behind decisions to migrate, it is worth briefly discussing the process of older people’s migration or retirement migration. Typologies of older people with migration experience are based on what stage of life, the age they experienced migration and where they are currently (King et al. 2021). The first group may be ageing migration pioneers, the first generations of migrants. A significant number of them have not returned to their country of origin after retirement and participate in various types of migration: they often circulate between other countries as needed. The second group consists of people undertaking migration at an older age. This group will include, for example, women who undertake caring work, e.g. for grandchildren in their own family abroad; or older migrants joining adult children of migrants. The third and most privileged group are older people looking for a better life, what is called lifestyle migration, retirement migration, amenity looking migration or environmental preference migration (Williams et al. 1997). This relates to individuals for whom the main motive of migration is not economic, but rather the quality of life and aesthetic considerations. It should be noted that here the decision to migrate is taken by people who are wealthy, independent and retired. This type of migration involves moving to a place set in agreeable rural surroundings or with a sunnier climate where a more pleasant and healthy lifestyle can be enjoyed. There are several variants of such migration. Some want to 'escape to the sun' and live in resorts on the Mediterranean coast of Spain, others are 'international counterurbanizers' such as people from the UK who buy homes in the French countryside, and 'countercultural migrants' like the Germans and Danes, who settle in Ireland to lead 'alternative' rural lifestyles (King 2002). The example of European migrants who spend their time in the sunny south well illustrates the fact that the boundary between migration and other forms of mobility is difficult to capture. The range of people movements includes tourist trips along with seasonal and permanent relocations to attractive areas. A relatively new phenomenon is migrating to provide care services in countries where care work is cheaper, e.g. the Philippines or Thailand. A relatively new phenomenon is migrating to provide care services in countries where care work is cheaper, e.g. the Philippines or Thailand. The migration of older men to these countries to find a younger partner who will become a caregiver in old age may be similar. This is an example of the reversal of the directions of marriage migration, which I wrote about above. The scale of these types of migration is difficult to determine on a global scale as we do not have data concerning the motives for migration, but it is, nevertheless, an interesting phenomenon that highlights the multidimensional nature of the migration process.

The issue of the demographic characteristic of migration flows is not new, but now, it is also raised, for example in Europe, in the context of an ageing population. International migrants often comprise larger proportions of working-age persons compared to the native population, and, depending on their country of origin, tend to have more children, these two elements can contribute to temporarily slowing the long-term trend towards population ageing in countries of destination. Population ageing underlies the increasing demand for migrant workers in many high-income destination countries, particularly for caregivers and healthcare female workers. Individual highly developed countries compete with each other to obtain the appropriate number of employees who are providing care services. In countries of origin, the emigration of the working-age population can help ease pressures on the labour market but also can increase the share of an older person in the immobile part of the population and even population shrinkage in some parts of the country.

2.5 The Scale of Forced Migration, Hosting Countries for the Refugees and Asylum Seekers

One of the most important subjects discussed in media today is the increasing scale of forced migration. There is a belief that most refugees come to Europe or the USA, which is more related to the political and media context than the actual scale of migration.

A forced migrant is a person subject to a migratory movement in which an element of coercion exists, including threats to life and livelihood, whether arising from natural or man-made causes (i.e. movements of refugees and internally displaced persons as well as people displaced by natural or environmental disasters, chemical or nuclear disasters, famine or development projects) (IOM 2019). What characterizes refugee migrations (migrations in search of international protection) is their cause, i.e. various forms of persecution by the state apparatus attempts to eliminate groups with different views (political or religious) or belonging to other nationality or ethnic groups. Refugee migrations may be triggered by the outbreak or escalation of wars. To protect people migrating for this reason, an international system of assistance to refugees was created (related to the widely known Geneva Convention). The convention remains one of the most popularly adopted international documents (146 countries signed it). In the text, I use the category of forced migration in a broad sense, regardless of recognition by individual countries or legal provisions.

Between 2000 and 2020, the number of persons displaced across international borders while fleeing conflict, persecution, violence or human rights violations doubled from 17 to 34 million. It represents about 16% of the total increase in the number of international migrants worldwide. In 2020 refugees and asylum seekers represented 12% of the global migrant stock, compared to 9.5% two decades earlier. Over 80% of the world's refugees and asylum seekers were hosted in low- and middle-income countries. Refugees and asylum seekers comprised only 3% of all international migrants in high-income countries compared to 25% in middle-income countries and 50% in low-income countries (UNDESA 2021).

The data show the duality of forced migration processes. Of course, it is difficult to distinguish between different types of migrants and most of the countries of destination benefit from the skills and fiscal contributions of labour migrants and refugees and asylum seekers but the countries predominantly gaining from the transformative power of labour migration are more often a high-income, while those shouldering a disproportionate responsibility for assisting to populations displaced across national borders as a result of conflict or persecution are often low- and middle-income countries (UNDESA 2021) (Table 2.2).

In 2020, one in five (or nearly 6.7 million) of all internationally displaced persons as a result of conflict or persecution was born in the Syrian Arab Republic. The second-largest number of refugees and asylum seekers globally came from the State of Palestine (5.7 million); equal to one in six of the world's total (UNRWA 2020; UNHCR 2020). The Bolivarian Republic of Venezuela was the origin of the third-largest number of internationally displaced persons, with over 90.000 refugees, nearly

Table 2.2 Countries of origin and host countries with the highest number of refugees in 2022

Countries of origin	Number (million)	Host countries	Number (million)
Syrian Arab Republic	6.8	Turkey	3.7
Venezuela	5.6	Colombia	2.5
Ukraine	5.4	Germany	2.2
Afghanistan	2.8	Pakistan	1.5
South Sudan	2.4	Uganda	1.5

Source UNHCR (2022a, b) (www.unhcr.org/refugee-statistics)

800.000 asylum-seekers and 3.6 million Venezuelans displaced abroad (UNHCR 2020). Turkey hosted the largest number of refugees and asylum seekers worldwide (nearly 4 million), followed by Jordan (3 million), the State of Palestine (2 million) and Colombia (1.8 million). Other major destinations of refugees, asylum seekers or other persons displaced abroad were Germany, Lebanon, Pakistan, Sudan, Uganda and the USA. These data indicate that the perception that most refugees come to Europe is not justified. It can be said that in 2015 Europe was confronted with its geographical location: the vicinity of the Middle East and the location by the Mediterranean Sea were influenced by migration caused inter alia by the war in Syria.

The unfortunate reality is that in the future probably we will observe a further increase in the number of forced movements. In the last few years, some major migration and displacement events have been seen. Foremost, the displacements of millions of people due to conflict (such as within and from the Syrian Arab Republic, Yemen, the Central African Republic, the Democratic Republic of the Congo and South Sudan), extreme violence (such as inflicted upon Rohingya forced to seek safety in Bangladesh) or severe economic and political instability (such as faced by millions of Venezuelans) have been seen. There has also been growing recognition of the effects of environmental and climate change on human mobility (such as planned migration/relocation and displacement), including as part of global efforts and international policy mechanisms to address the broader impacts of climate change. Large-scale displacement generated by climate and weather-related hazards occurred in many parts of the world in 2018 and 2019, including in Mozambique, the Philippines, China, India and the USA.

To summarize, it should be noted that the number of forced migrants has been increasing for many years. They are of a varied nature and are both internal and international, triggered by both conflicts and natural disasters. The protracted conflicts and the growing importance of climate change may indicate that the number of forced migrants will continue to increase in the coming years. An interesting issue is the concentration of migrants in lower and middle-income countries and the policies of important immigration countries, such as the US or the EU, which try to keep the largest possible group of forced migrants abroad through agreements with the so-called third countries (i.e. the US-Mexico agreement or the EU-Turkey, EU-Morocco). As a consequence, a select few people who are recognized as refugees under this system receive a privileged status that guarantees access to valuable resources such as the right to reside and the ability to use public services in social security countries. The process of granting international protection itself has become a tool for segregating refugees and de facto limiting the movement of those considered undesirable (FitzGerlad 2019). The international refugee regime, which was created to protect refugees, is gradually offering them less and less help and security and increasingly restricting access to aid. On a global scale, the poorest regions of the world provide shelter to refugees to the greatest extent, and thus disproportionately struggle with the consequences of actions resulting from the (colonial) policies of the richest countries in the world.

2.6 The Influence of the COVID-19 Pandemic on the International Migration

Research on migration indicates that the year 2020 was an important turning point in human mobility. We could observe a significant reduction in movements of all kinds, especially connected with tourism and business travel, labour migration or those related to international student's mobility or family reunification. The first travel restrictions were introduced in January 2020, and their scale has changed over time throughout 2021. In the early phase of 2020, countries introduced very preventive restrictions and health requirements, and many fully closed the possibilities of entry or prohibited travel from affected regions. Mobility was curtailed especially from March to May as populations were forced to stay home under national lockdowns. The number of passengers on international flights in April and May was down by 92% relative to the same months in 2019 (Benton et al. 2021). As it became clear that emergency measures were likely to continue for longer than originally planned, many governments sought to minimize the worst economic, and especially human, cost. Some countries introduced exceptions to the mobility ban, including nationals and residents (and their family members), diplomats and staff of international organizations, and healthcare workers. During the second phase, which fell during the summer period, countries decided to reopen some points of entry and bans were replaced by health measures like health declaration forms, pre-departure COVID-19 tests or quarantine measures. Until September different strategies crystallize. Some countries open up to tourists (like Caribbean islands) while others maintain border closures (i.e. New Zealand). According to Benton et al. (2021), much of Asia continued to employ route restrictions throughout most of this phase, even while adding on additional health requirements for travellers, and some Asian countries, including Japan, the Republic of Korea and Viet Nam, used their visa systems to limit travel to a larger degree than countries in other regions. In Europe, additional health requirements were most often used to reduce travel. The third phase can be called a phase of responses to new outbreaks and virus mutation. From September 2020 until January 2021 the situation was mixed. Some countries were trying to build their capacity to operationalize health measures in place of travel restrictions while battling a second (and in some cases, third) wave of infections and grappling with the emergence of new variants of the virus. Some countries, including Chile, Mexico and the United Arab Emirates, opened even to tourists. The fourth phase is connected with the first quarter of 2021 when a vaccine was developed and vaccination for COVID-19 started. The phase is not associated with fewer restrictions. New variants of the virus identified, i.e. in Brazil, and India caused, that it was expected that there is a long way from removing the restrictions. In most countries, travellers for a long time entered a country had to present proof of a negative COVID-19 test taken within 72 h of departure or documents of recovery from COVID-19. Meanwhile, the European Union was discussing ways of introducing the EU COVID certificate, the so-called COVID passport, which allowed vaccinated people to travel more freely, but also ways to adjust its risk ranking system to allow for stricter measures for the highest-risk areas. The epidemic situation in the world is stabilizing, and mobility has returned to normal, but it is not known when and if some new factor will not disrupt this state.

The COVID-19 pandemic shows mobility injustice as it influenced the mobility prospects of some groups while making a smaller difference to others. Those whose resources or nationality and status enable them continued mobility for work, family or tourism, but those who move out of necessity (such as migrant workers or refugees) were stopped because of the lack of possibility to absorb expansive quarantine and self-isolation costs (Benton et al. 2021). Looking ahead the gap between movers and non-movers can widen, especially if people with resources and opportunities move freely and restrictions related to COVID-19 or pre-existing visas, and travel restrictions are implemented to limit the migration of disadvantaged groups. It can be the case also when travel begins to favour those who have been vaccinated or tested or if reliance on digital health records makes a person's ability dependent on their digital access and literacy.

The pandemic strengthened the socioeconomic vulnerability, of those who depended on mobility for survival. Job losses hit migrant workers hard, and national lockdowns did not allow some labour migrants to depart and some to arrive at a given place. They were also hit by social safety nets which usually were minimally available for them. Restrictions increase the dependence of migrants on intermediaries and facilitators, from employment agencies to smugglers. This was especially due to difficulties in access to reliable information about fast-changing migration routes. Moreover, travel restrictions have increased the demand for smuggling services among people desperate to flee economic deprivation, natural disasters or violence as well as people attempting to return home. The closure of borders and more restrictions have pushed smugglers to use more dangerous routes and raise their prices—exposing migrants and refugees to an increased risk of exploitation and trafficking (Benton et al. 2021).

We do not know how this or subsequent pandemics will influence migration. In 2022, governments faced further challenges in developing risk mitigation strategies that move beyond the ineffective tools of border closures and travel bans. They needed to avoid one-sided responses and work with other governments and international organizations to develop health policies that were well-planned and promulgated. The solution and strategies that will be built now, will be with us not only for this but also for the next pandemic, and it is thus even more important whether the system both minimizes public health risks and is inclusive of all people on the move. The second element of inclusivity seems crucial in forming the perspective of mobility justice.

2.7 The Transformation of Europe from a Region of Emigration into Predominantly a Region of Immigration

Apart from identifying the most important features of migration on a global scale, it is worth looking at changes at the regional level. To show the processes taking place in more detail, I would like to present the changes taking place in Europe. Europe, formerly the main source of emigration, has become the most important reception area and the changes in the approach to migration and migration policies will be described in more detail in this chapter. Europe is also a good example for discussing interregional migration, as in 2020 it had the largest share of intra-regional migration, with 70% of all migrants born in Europe residing in another European country. The pattern of labour migration within the EU and changes in integration policies will also be presented. The last part describes new characteristic types of migration flows in Europe, such as a migration of knowledge, fluid migration or spill-over of migration flows.

In the first years after World War II, Europe was primarily an emigration area. Central and Eastern European countries lost many citizens who ended up in the West as displaced and refugees, and the inhabitants of Western Europe emigrated mainly to the USA, Canada, Australia, New Zealand and Latin America. The word 'emigration' then meant migration to settle elsewhere. The situation began to change in the late 1950s and early 1960s. Starting from Western Europe, emigration decreased and immigration grew in importance. Western Europe has become a continent of immigration, although some European countries have consistently been defined as non-immigrant countries. Defining as non-immigrant had consequences for the policy pursued, as well as for the perception and description of the phenomenon of immigration. In many countries, the influx of newcomers was legitimate, but not equated with immigration. For example, in the post-war period in the UK, people were allowed to come and work under the British subject, including from the areas of the former British colonies. In the Netherlands, residents of the former Dutch East Indies, when Indonesia gained independence in 1945, were recognized as repatriates, migrants from Suriname (Dutch Guiana) and the Netherlands Antilles as overseas citizens and were also allowed to come. In Germany, the same situation was with the admission of so-called ethnic Germans (Aussiedler). Additionally, in Germany, an important group of migrants were guest workers (Gastarbeiter), which suggested only temporary stay. After the oil crisis in 1973, the possibility of arrival was limited, especially for economic migrants. The closing of the door to new employees increased the number of people migrating as part of the family reunification process. Since the mid-1980s, the number of asylum seekers increased in Western Europe and the culmination of this migrant wave in the first half of the 1990s led to the first migration crisis. The crisis contributed to tightening the controls and criteria for admitting newcomers. The new regulations increased the scale of illegal attempts to enter Europe and migration began to be perceived as a source of problems and threats.

The perception of migration from the point of view of a 'non-immigrant country' in the past also influenced the shape of the national integration policy. Some categories of migrants were treated a priori as members of society (such as the aforementioned British Subjects, Dutch repatriates, or ethnic Germans) and were offered full citizenship and assimilation programs. For workers who were treated as temporary visitors, limited, usually ad hoc, integration measures were offered, mostly in the socio-economic dimension, maintaining the illusion that migrants would return to where they had come from.

For many years there was a significant variation in approaches to integration and integration policy in European countries. The differences resulted not only from the different ways of interpreting the concept of 'integration' but also, inter alia, from political and ideological preferences, various mechanisms and instruments of politics, the history of immigration or the social situation of the immigrants themselves. In the literature, we can find numerous attempts to create models to explain past differences in immigrant integration policies. An interesting division was used, for example, by Carrera (2005). He distinguished countries representing a multicultural, assimilative and excluding model. The UK was a European example of a multicultural model where the coexistence of cultures was visible, but as separate communities, functioning side by side, but not striving for the state of homogeneity. The challenge in this model was to enable the harmonious coexistence of these communities in a multicultural state, hence the integration activities were directed not at individual people but at minority groups. The European 'assimilation type' was France. The strategy of assimilation assumed the inclusion of foreigners in the native society, migrants should adopt the applicable customs and respect constitutional rights and obligations. Immigrants were perceived as individuals, hence integration concerns individual people. The disadvantage of this model was the assumption of a one-sided process of the migrant's adaptation to the host society and the recognition that this process would occur spontaneously. The exclusion model based on the principle of *ius sanguinis* (the law of blood) is no longer encountered in modern Europe. However, it was carried out almost until the end of the twentieth century in Germany and Belgium. This model assumed the least possible integration or, in an extreme form, segregation of immigrants. This model was characteristic of countries that considered migration to be only a temporary process, hence there was no reason to include migrants in integration programs. Moreover, it assumed that integration could prevent a migrant from returning to his country. This model was based on erroneous assumptions and is therefore no longer implemented. Denying the existence of migration has delayed the integration process and created divided communities, living side by side but not together.

Starting in 1990, patterns of migration from, towards, and within Europe changed significantly and went through further diversification. New migration flows across Europe were introduced by the collapse of the Iron Curtain and the opening of the borders of Eastern Europe. Since the 1990s, the European Union has become a new, active player creating policies that affect international mobility and migration in Europe. The 1992 Maastricht Treaty's abolition of borders considerably eased intra-EU movements, at the same time, entrance into the EU became progressively restricted due to the unification of the European market, which imposed strict border controls and visa regulations (Van Mol and de Valk 2016). Migrants' countries of origin as well as their migration motives became increasingly diversified. EU member states changed their national policies into common, restrictive and defensive the entry of potential migrants from outside the EU. The institutional framework for the regulation of migration and integration, created mainly by Western European countries, the first EU members, became a model for most EU countries, and also for the new member states, which are required to legislate under the existing EU policy. In reality, however, there is a discrepancy in the visions of shaping migration policy between the western, southern and eastern countries.

Migration policies of individual countries evolve under the influence of both internal factors (i.e. the needs of the labour market) and external factors (i.e. non-European conflicts). The recent financial and economic crisis of the end of the first decade of the twenty-first century and the so-called migration crisis of 2015 strengthened anti-immigration sentiment. In the coming years, we can expect a further tightening of the migration policy and the maintenance of anti-immigration rhetoric. Short-term political interests and a sense of insecurity—often expressed as a cultural and religious issue—prevent the formulation of a long-term, proactive migration policy. It should be noted that although (officially) the migration policy is restrictive, there is a constant influx of various groups of migrants, also from non-European countries, the number of migrants is growing and this trend is unlikely to change.

European societies' attitudes towards immigrants and philosophies of integration are constantly evolving under the influence of changing social realities, and political and economic priorities.[5] The problems which are currently being experienced by host countries stem from the failure of a large number of the policies implemented thus far. One of the many examples of this is the creation of 'parallel societies' where the representatives of different nationalities live next to each other and not together. Governments are faced with the challenge of how to ensure social cohesion in a context of extreme social, ethnic and cultural diversity. Numerous tensions between 'traditional' communities in the host countries and first, second and even third-generation immigrants have found their outlet in a growing number of public disturbances (i.e. in the suburbs of Paris), assaults (especially on Muslims) and even terrorist acts. There is a belief that migration is a constant and positive process, and states are evolving towards multicultural societies.

The integration policy changed only in the first decade of the twenty-first century, but it mainly still concerned Western Europe. Countries have moved from previous

[5] Documents of international institutions (such as the International Organization for Migration and the European Commission) define integration as a two-way process by which immigrants become an accepted part of society. Integration assumes that migrants and hosts will respect all existing laws and obligations as well as values linking migrants and the host society (Carrera and Atger 2011). The recent documents indicate that integration is a three-way process where also the sending country should be included (Garcés-Mascareñas and Penninx 2016). Although there is agreement as to the validity of understanding integration as a process, there are somewhat large differences as to the depth of the adjustment to be made by sides and the effect to be achieved (i.e., whether the principle of multiculturalism should prevail or not).

rights-based integration concepts to cultural-sensitive integration that focuses on shared cultural values—necessary for social cohesion. Greater attention to cultural issues resulted from how the host societies defined their identity (i.e. modern, liberal, democratic, equal, etc.). In the years 2001–2009, we could see an increase in the number of projects addressed directly to migrants, the integration policy was made more specific, and the number of non-governmental institutions dealing with this issue was expanded. In the following years, for example in the EU countries, we could observe a tendency to include integration policy in all local policy activities, the so-called mainstreaming integration policy. Resigning from dedicated programs does not mean that the network of organizations supporting integration has been liquidated. The best solution is, of course, to combine these two approaches and conduct dedicated activities in areas where general measures cannot be taken, although they are difficult to implement due to limited financial resources. Recent years brought changes in the approach to immigration and integration as a consequence of the economic and migration crisis. The policy is aimed at limiting the inflow from outside (the so-called migration management, assuming the admission of people important from an economic point of view, was introduced) and focusing on solving the problems of groups who already live in Europe (hence such a large number of third-generation immigrants) (Jaczewska 2015). Most European countries developed national recommendations for the implementation of integration policy and delegated the competencies to conduct activities at the local level—to local authorities and non-governmental organizations. Currently, the role of local communities is emphasized and they are encouraged to shape a bottom-up integration policy, as integration takes place in a local environment. Experience related to politics shows that local communities are best able to respond to changing needs and create programs in the best way (Jaczewska 2013). In European countries with extensive migration experience, we deal with a whole network of non-governmental institutions that cooperate with local authorities by carrying out integration activities. In these countries, organizations were established not only in the largest cities but also in medium-sized cities, wherever there was a demand for their work. The important assumption is that integration policy must be conducted comprehensively, i.e. it must include all dimensions of integration: the socio-economic, spatial, cultural, legal and institutional dimensions. Today, most activities in all of Europe are created under the first dimension. It is extremely important from the point of view of the foreigner's ability to function independently. However, other dimensions should not be forgotten. After the migration crisis, we can observe a new trend in the development of integration policy which can be called the whole-of-society approach (Papademetrious and Benton 2016). It signifies that integration is not only the responsibility of agencies dealing with migrants, and non-governmental organizations focused on helping newcomers, but the entire society. It means engaging multi-sectoral stakeholders and facilitating their active participation in the decision-making process to take appropriate measures together. The Whole-of-Society Approach derives from tactics to prevent multi-hazard risk but also help implement integration programs. Integration policy has become the 'political slogan', but it also remains a significant challenge for most countries that have accepted, are accepting and intend to accept migrants and refugees.

Nowadays, the largest number of migrants in Europe is accepted by the EU and its associated countries. According to Eurostat (2021) data, in 2020, 23 million people with citizenship of the so-called third countries (non-EU), accounted for 5.1% of the population of EU countries. 13.5 million persons were living in one of the EU Member States with the citizenship of another EU Member State, which shows the scale of intra-EU migration. The largest number of non-nationals living in the UE Member States was in Germany (10.4 million), Spain (5.2 million), France (5.1 million) and Italy (5.0 million). About 71% of non-nationals in the entire EU live in these four countries. In recent years, Spain has become a country with the second-largest population of immigrants, because of a stable inflow of new migrants. The UK dropped out of the ranking after Brexit but was previously the second EU country with the largest number of immigrants. In relative terms, Luxembourg was the EU Member State with the highest percentage of non-nationals, as non-nationals accounted for 48% of the total population. A high proportion of non-nationals (10% or more of the living population) has also been observed in Cyprus, Austria, Estonia, Malta, Latvia, Belgium, Ireland and Germany. On the other hand, non-nationals (before the war in Ukraine) accounted for less than 1% of the population in Poland (0.9%) and Romania (0.7%) and Lithuania (0.9%). Belgium, Ireland, Luxembourg, Austria, and Slovakia were the EU Member States whereas non-nationals were mainly from other Member States. This means that in those EU Member States, non-nationals were citizens of countries inside the EU. In most countries, we could see an increase in the number of non-nationals of non-EU countries. Some of these increase is related to the inflow of undocumented migrants, mostly from Africa (i.e. in Spain and Greece), some are connected with the development of new migration possibilities created in former immigrant countries (i.e. in Poland), and some are a consequence of the new programs that attract high-skilled workers or students from all around the world (i.e. in Germany).

Nearly 60% of all non-citizens war people from the UE, that reside in another EU country, in 2020. The two most important directions of intra-EU migration can be distinguished. The first one was from the east to the west, or, more precisely—from the new EU countries (EU10 + 2 + 1) to the old countries and the second one was between neighbourhood countries. In the years 2004–2019, the number of people coming from the countries that joined the EU in 2004 and 2011 and residing in the EU-15 countries increased significantly and the citizens of Romania and Poland had the largest share in this growth. The neighbourly nature of migration was noticeable in Luxembourg (migrants from France, Belgium, and Germany), Finland (from Russia, Sweden, and Estonia), Austria (from Germany, and Hungary), Belgium (from France, and the Netherlands), Sweden (from Denmark, Finland), Germany (from Poland), the Netherlands (from Germany), Portugal (from Spain) and Denmark (from Germany), as well as (outside the EU) Switzerland (from Germany, France and Italy) and Norway (from Sweden).

The trends in inter-EU labour migration show large differences between regions and countries in Europe. The regions of Eastern and Southern Europe have a substantially negative net migration, while the regions of Northern and Western Europe show a positive balance. However, even in countries hosting large numbers of migrants, new areas of the internal periphery are emerging. ESPON's (2020) research indicates the existence of four types of the inner periphery: (1) areas with high travel times to regional centres; (2) areas with low economic potential; (3) areas of poor access to public services; (4) impoverishing areas with low levels of economic and demographic indicators. The main factors driving labour migration are the conditions and situation in the labour market. European development is increasingly based on the knowledge economy. The presence or absence of a highly developed service sector becomes the dominant factor explaining the direction of employee migration. Foreign workers are more often employed in metro regions than in other types of regions. Moreover, highly educated migrants tend to settle in regions where residents are also well-educated—which further exacerbates intra-regional disparities. ESPON (2020) research developed several regional typologies to describe the main characteristics of regions and cities in terms of attracting immigrants. It was recognized that migration attractiveness is influenced by economic, demographic and social conditions, including the quality of governance, public services, integration policy and the local political and social climate. Southern and Eastern Europe is shown as the least attractive area for migrant workers, which may be the basis for forecasting that migrants will still be more willing to migrate to a highly innovative part of Western Europe.

Modern intra-EU migrants seem also to be adopting 'liquid migration strategies' (Black et al. 2010), which means that we are very mobile and it is hard to predict what the movements will look like soon. Today, migration destinations are subject to very rapid change, and active choices are made to move to where opportunities are available at a given moment. Migrants are reacting very quickly to changes, especially those who are political and economic migrants. There are also new forms of mobility that we can define as knowledge migration. For modern organizations, the mobility of workers is not only the physical migration of people but also the transfer and exchange of individual knowledge. This means travelling on business and attending conferences and meetings that take place in multiple locations. Migration is also not seen as a one-time move to settle in a new place permanently, but it takes the form of a fluid migration based on the temporariness, flexibility and unpredictability of life trajectories. Highly qualified employees of large commercial organizations or the academic sector are predisposed to this type of mobility. Another process, which is currently visible, is the 'spillover of migration' into ever-larger areas both on a European scale and at the level of regions and countries. Inter-EU migrants are no longer choosing as destinations only the largest cities and central areas, but they are travelling to places previously untouched by migration inflows (i.e. small- and medium-sized towns, as well as peripheral centres) but where there is now a need for their work.

European countries are experiencing also migration from outside the EU, which is it is perceived differently from inter-EU migration. The most visibly discussed are aspects of policies implemented towards refugees. The policy towards refugees has always been present in European countries, but it has acquired a new meaning in the face of increased migration, which took place under the so-called 'European migration crisis' in 2015. Researchers of the International Association for the Study of Forced Migration (IASFM) and activists of organizations supporting refugees point out the unfortunate nature of this term and postulate (which I agree with) that we should talk about a crisis of humanitarian protection or a crisis of the European Union as such. The humanitarian protection crisis has been described as the sharp increase in the number of migrants coming to Europe via the Mediterranean Sea and south-eastern Europe to obtain asylum. Although the number of migrants has been increasing for a long time, the crisis is considered to have started in 2015, when a record 1.3 million asylum applications were submitted in the EU countries. The crisis should be seen as part of the global humanitarian crisis. The wave of social unrest that started in Tunisia in December 2010 and spread to North Africa and South-West Asia (the so-called Arab Spring) was extremely important from the European point of view. Tensions in Tunisia and Egypt have had relatively limited migratory effects, but the conflicts in Libya and Syria have generated a significant flow of refugees. In 2015, over a million people tried to cross the Mediterranean Sea to the EU, including 848 thousand who came to Greece, and 153 thousand to Italy (at least 3.735 people were killed). Among the refugees and migrants arriving by sea in 2015, 58% were men, 17% were women and 25% were children. In 2015, the most popular route was the Western Balkans, leading through Turkey, the Aegean Sea, Greece (mainly the islands of Kos, Chios, Lesbos and Samos), Macedonia and Serbia to Hungary. After the Hungarian-Serbian border was closed in September 2015, the road changed, leading from Serbia to Croatia. The West African, West Mediterranean and Central Mediterranean routes, known for many years, were less important in this period. In 2015, the largest number of asylum applications were submitted in Hungary, Sweden and Austria. In absolute numbers, Germany was the most popular (476 thousand). The number of asylum applications submitted by Syrian refugees in Europe reached 383 thousand in 2015 (the largest number of Syrians were and are still being hosted by neighbouring countries such as Turkey, Lebanon and Jordan). The number of applications from the Western Balkans (mainly Kosovo and Albania) reached 201 thousand, Afghans applied 196 thousand applications (UNHCR 2015; Jaczewska et al. 2018).

The European humanitarian crisis made politicians realize not only the scale of potential migration to Europe, and the determination of migrants to apply for asylum but also the lack of unanimity and solidarity in the European Union and the inability to deal with the crisis, increasing social unrest and anti-immigration sentiment. The European Union has been implementing the strategy developed under the European Migration Agenda since 2015 and is trying to limit (or stop) the flow of migrants by increasing funding for border patrol operations in the Mediterranean Sea; developing plans to combat migrant smuggling; proposing a new quota system both to relocate asylum seekers in the EU among European countries, to ease the burden on countries on the external borders of the Union, and to resettle people who have been recognized as genuine refugees; and providing assistance to refugees residing outside the EU. An important step that contributed to reducing the number of refugees reaching Europe was the development of cooperation with third countries—especially the agreement with Turkey in 2016. The number of asylum seekers decreased compared to 2015, reaching the value of 647.000 in 2018. In general, efforts have focused on protecting external borders, including the maritime border (Crawley et al. 2018). People arriving in Europe are still directed to overcrowded temporary 'hotspots' (camps), an example is one in Lampedusa. In 2018, the European Commission published a progress report on the implementation of the European program, analysing the progress made and the shortcomings in its implementation. An important conclusion was the recognition that climate change and demographic and economic factors create new reasons forcing people to move, which shows that the migratory pressure has not disappeared and that it is necessary to keep the migration situation under control. In March 2019, the European Commission declared an end to the migration crisis, and European migration programs are still being introduced (European Commission 2020).

The asylum recognition rates in the EU before and after the crisis, between 2008 and 2017 varied. The highest was observed in Finland, Ireland, Switzerland, and the Baltic States. The highest average recognition rate was for migrants from Syria, Yemen and Eritrea (Map and graph).

In 2020 in Europe, asylum seekers came from nearly 150 countries. The number of applicants was estimated at 472 thousand, including 417 thousand of first-time applications. A growing share of applicants come from visa-free countries who enter the EU legally, mostly from: Venezuela, Colombia, Georgia and Peru. These are new geographical directions from which in 2020 applicants ask for asylum. There is still a large number of pending applicants compared to the total number of applications in a given time and there are significant differences in rates of recognition. Overall, EU countries granted some sort of protection to around 280,000 people in 2020. The largest groups were from Syria (27% of all people granted protection), Venezuela (17%) and Afghanistan (15%) (Eurostat 2021).

The year 2022 created another challenge connected with an unexpected wave of people forced to leave their homes because of Russian aggression towards Ukraine. Russia's invasion of Ukraine has led to the largest migration and related humanitarian crisis in Europe since World War II. According to UNHCR data from May 2022, more than 6.8 million people[6] have left Ukraine since February 24. At the same time, 2.3 million Ukrainian citizens returned to Ukraine in the same period (UNHCR 2022a, b). It is assumed that by the end of May, there were about 4.5 million war refugees from Ukraine outside the country and Poland became the country with the largest number of border crossings and the number of remaining war refugees from Ukraine (Polish Border Guard 2022). It is worth noting that from the first days of the war, civil society, including volunteers, social organizations, formal and informal groups and the Ukrainian diaspora, came to help people fleeing from Ukraine, responding to the most urgent needs, including transport from the border, providing housing and food.

According to data from the Union of Polish Metropolises from the turn of May and April 2022, as many as 67% of adult Polish women and men are involved

[6] Ukrainians and foreigners were staying in Ukraine when the war broke out.

in assisting refugees. In response to the crisis, international organizations such as UNHCR and IOM, as well as local governments were trying to professionalize the work of volunteers and quickly intervened. The Polish government developed a legal framework for assistance, and it was loosely based on EU regulations (Directive 2002/33/EC) introducing temporary protection. The aid provided in Poland was broader than that required by the European Union, but the list of people supported was narrower. Both in Poland, and in European countries, the assistance was faster than in the case of refugee granting procedures, it provided quick access to social assistance on conditions similar to those of people with Polish citizenship, but it should be noted that it provided weaker legal protection (than for refugees) because it did not guarantees access to integration programs or a path to citizenship.

According to Duszczyk and Kaczmarczyk (2022), it is very difficult to predict whether migrants will decide to stay in Poland, go to other countries, or return to Ukraine. The key to the decision may be the circumstances accompanying the migration decision, i.e. having dependent family members, education, or the ability to adapt to a new country. The resistance of the Ukrainian army and society to Russia's aggression shows that the scenario of a quick end of the war is very unlikely. Long-term conflict will cause continuous migration of refugees. This massive inflow of foreigners in a short period causes the simultaneous occurrence of various challenges related to their stay and integration into the society of the host country. The most important and the most difficult to solve are those related to housing, education, health care, and the labour market. Apart from creating a good legal framework ensuring access to many forms of assistance (social assistance, health services and the labour market), the government was not very involved in crisis management. By November 2023, it had not created a long-term strategic and integration framework. Most of the activities rested on the shoulders of social and local government organizations. Based on the experience from 2015, a strong expansion of refugee protection may create several threats. The most important of them include, e.g. unstable financing or underfinancing, a significant number of short-term activities carried out in the project system, uneven quality of services provided and poor coordination of activities.

2.8 Summary and Conclusions

This part of the book aimed to describe challenges connected with global international migration. Firstly, global drivers and factors influencing migration were described. Author pointed out five aspects, which are connected with economic, social, political, technological transformation, and environmental changes, and tried to demonstrate that these factors create interconnected conditions, which are frames for the complex, variable international migration movements. The contextual and processual nature of migration were highlighted. In the following parts, special emphasis was put on characteristic features of a contemporary international movement, the scale and direction of flows and the significance of age or gender structure. Moreover, the scale of forced migration and the influence of COVID-19 were outlined. The chapter relating to the main global directions and forms of migration is inevitably rather brief and superficial.

Secondly, author characterized Europe as a region that transformed from predominant emigration to immigration. Here specially Author described the changes in migration and integration policies and the changing nature of the migration movement. The last part described two migration humanitarian protection crises that influenced and will influence further migration and the development of policies related to international migration. Shortly, the situation of the crisis after the outbreak of the conflict in Ukraine in 2022 was discussed. Recent crises have highlighted the challenges associated with the functioning of the refugee protection system in Europe. Public discourse has intensified, and various threats posed by migration are increasingly raised. This results in difficulties or denial to people and, consequently, in the illegal expulsion of migrants or push-backs. There are also different responses to the arrival of divergent groups of migrants; help is refused to migrants of a specific origin or religion and solidarity and help are declared to the other groups. Migration and migration-related diversity are likely to remain key topics of the European policy and research agenda for the foreseeable future.

This part of the book by no means exhausts the topic of international migration. Focusing on selected issues, author's intention is to counter stereotypical views on contemporary international migration and the migration situation in Europe. The phenomenon of intensified migration now covers all societies and is generally intensified in places where it has been relatively weak so far. It is difficult to determine what migration trends will look like in the future. Some regions of the world will experience a migration transition and become or consolidate their status as a net migration area (e.g. Mexico or Poland). Citizens in high-income countries will continue to choose lifestyle migration (especially baby boomers moving to countries with milder climates or lower costs of living). Refugees will probably continue, we will experience a greater degree of climate refugees, and this may lead to the introduction of further restrictions in migration policies. It should be noted that some trends cannot be predicted, such as forced immobility related to the COVID-19 pandemic or the outbreak of war in Europe.

Migration should not be perceived as a threat to state security but as a result of the human insecurity that arises through global inequality. People have always migrated to improve their livelihoods and to ensure greater security. International economic disparities, poverty and environmental degradation, combined with the absence of peace and security and human rights violations are all factors affecting international migration. A fundamental change in attitudes, also caused by unexpected events, can be an important step towards more fair and effective migration policies. Migration is an important aspect of human development and mobility is a basic freedom. It has the potential to lead to greater human capabilities and is a form of adaptation. Ensuring that people can move safely and legally helps enhance human rights, and also can lead to greater economic efficiency and social equality (UNDP 2009). This human development approach could provide a new frame of reference for thinking about migration and diversity.

References

Adhikari R et al (2014) The impact of parental migration on the mental health of children left behind. J Immigr Minor Health 16:781–789
Amelina A, Horvath K, Meeus B (eds) (2016) An anthology of migration and social transformation. European Perspective. Springer
Antman FM (2015) Gender discrimination in the allocation of migrant household resources. J Popul Econ 28:565–592
Asiedu E, Chimbar N (2020) Impact of remittances on male and female labour force participation patterns in Africa: quasi-experimental evidence from Ghana. Rev Develop Econ 24(3):1009–1026
Barone G, Mocetti S (2011) With a little help from abroad: the effect of low-skilled immigration on the female labour supply. Labour Econ 18(5):664–675
Benton M, Betalova J, Davidoff-Gore S, Schmidt T (2021) Covid-19 and the state of global mobility in 2020. https://publications.iom.int/books/covid-19-and-state-global-mobility-2020
Black R, Engberson G, Okólski M, Panţîru C (2010) Continent moving west? EU enlargement and labour migration from Central and Eastern Europe. IMISCOE Research, Amsterdam University Press
Boeri TJ, Conde-Ruiz I, Galasso V (2012) The political economy of flexicurity. J Eur Econ Assoc 10(4):684–715. https://doi.org/10.1111/j.1542-4774.2012.01065.x
Bryceson DF (2019) Transnational families negotiating migration and care life cycles across nation-state borders. J Ethnic Migr Stud 45(16):3042–3064. https://doi.org/10.1080/1369183X.2018.1547017
Caponio T, Cappiali TM (2016) Exploring the current migration/integration 'crisis': what bottom-up solutions? Vision Europe Summit. In: Migration and the preservation of the European Ideal, Calouste Gulbenkian Foundation, Lisbon. https://gulbenkian.pt/wp-content/uploads/2016/10/ExploringtheCurrentMigration_IntegrationCrisis2016.pdf
Carrera S, Atgar AF (2011) Integration as a two-way process in the EU. In: Assessing the relationship between the European integration fund and the common basic principles on integration. https://ec.europa.eu/migrant-integration/?action=media.download&uuid=2A7D3EAF-9701-AB1A-B2070276212EABA5
Carrera S (2005) Integration as a process of inclusion for migrants? A case of long-term residents in the EU. CEPS Working Document 2019, Brussel
Castles S (2013) The forces driving global migration. J Intercult Stud 34(2):122–140. https://doi.org/10.1080/07256868.2013.781916
Cho R (2019) How climate change impacts the economy. State of the Planet. Available at: https://news.climate.columbia.edu/2019/06/20/climate-change-economyimpacts/ (27.06.2021)
Cortés P, Tessada J (2011) Low-skilled immigration and the labour supply of highly skilled women. Am Econ J Appl Econ 3(3):88–123
Crawley H, Duvell F, Jones K, Macmahon S, Sigona N (2018) Unravelling Europe's migration crisis: journeys over land and sea. Policy Press, Bristol, United Kingdom
Czaika M, de Haas H (2013) The Effectiveness of Immigration Policies. Population Develop Rev 39(3). https://www.jstor.org/stable/23655336?seq=2#metadata_info_tab_contents
Czaika M, Parsons CR (2017) The gravity of high skilled migration policies. Demography 54:603–630
De Haas H, Czika M, Flahaux ML, Mahendra E, Natter K, Vezzoli S, Villares-Varela M (2019) International migration: trends, determinants and policy effects. Popul Dev Rev 45(4):885–922
Dolbin-MacNab ML, Yancura LA (2018) International perspectives on grandparents raising grandchildren: Contextual considerations for advancing global discourse. Int J Aging Hum Develop 86(1):3–33
Donato KM, Gabaccia D (eds) (2015) Gender and international migration: from the slavery era to the global age. Russel Sage Foundation, New York

Duszczyk M, Kaczmarczyk P (2022) Poland and war refugees from Ukraine—beyond pure aid. CESifo Forum, ISSN 2190-717X, ifo Institut—Leibniz-Institut für Wirtschaftsforschung an der Universität München. München 23(4):36–40

EIGE (2020) Gender Mainsteaming Sectoral Brief: Gender and Migration. https://eige.europa.eu/gender-mainstreaming (13.09.2022)

ESPON (2020) Addressing labour migration challenges in Europe. In: An enhanced functional approach policy brief. https://www.espon.eu/labour-migration

Ethical Journalism Network (2017) https://ethicaljournalismnetwork.org/media-and-migration (13.09.2022)

European Commission (2020). https://ec.europa.eu/home-affairs/what-we-do/policies/new-pact-on-migration-and-asylum_en

Eurostat (2021) Statistics on migration in Europe. https://ec.europa.eu

EUROSTAT (2022) Statistics on migration in Europe. https://ec.europa.eu

Farris SR (2015) Migrants' regular army of labour: gender dimensions of the impact of the global economic crisis on migrant labour in Western Europe. Sociol Rev 63(1):121–143

Ferrant G, Tuccio M (2015) South–South migration and discrimination against women in social institutions: a two-way relationship, world development. Elsevier, 72(C), pp 240–254. https://doi.org/10.1016/j.worlddev.2015.03.00

FitzGerald DS (2019) Refuge beyond reach: how rich democracies repel asylum seekers. Oxford University Press, Oxford

Garcés-Mascareñas B, Penninx R (2016) Introduction: integration as a three-way process approach? In: Garcés-Mascareñas B, Penninx R (eds) Integration processes and policies in Europe. IMISCOE Research Series. Springer, Cham. https://doi.org/10.1007/978-3-319-21674-4_1

Grabowska-Lusińska I, Jaźwińska-Matylska E (2013) Znaczenie migracji w życiu zawodowym kobiet i mężczyzn. Kultura i Społeczeństwo. PAN, nr 3. https://doi.org/10.2478/kultura-2013-0024

De Haas H, Castles S, Miller MJ (2019) The age of Migration. In: International population movement in the modern world. Macmillan Education

Hennebry JL, Petrozziello AJ (2019) Closing the gap? Gender and the global compacts for migration and refugees. Int Migr 57(6):115–138

Huang S, Yeoh BSA, Lam AMT (2008) Asian transnational families in transition: the liminality of simultaneity. Int Migr 46(4):3–13. https://doi.org/10.1111/j.1468-2435.2008.00469

Huysmans J (2006) The politics of insecurity: fear, migration and asylum in the EU. Routledge, London, New York

IOM (2019) Glossary on migration. Int Migr Law 34. https://publications.iom.int/system/files/pdf/iml_34_glossary.pdf

Jaczewska B (2013) Multidimensionality of immigrant integration policy at the local level: examples of initiatives in Germany and the United Kingdom. Miscellanea Geographica 1(17):25–33

Jaczewska B, Wites T, Solarz MW, Jędrusik M, Wojtaszczyk M (2018) Geographies of world population: demographic trends in the contemporary world. In: Solarz MW (ed) New geographies of the globalized world. Routledge Studies in Human Geography

Jaczewska B (2015) Zarządzanie migracja w Niemczech i Wielkiej Brytani. Polityka integracyjna na poziomie ponadnarodowym, narodowym i lokalnym. Faculty of Geography and Regional Studies, Univesity of Warsaw, Warszawa

Jaczewska B (2023) Analiza praktyk czasoprzestrzennych i sposobów zamieszkiwania „wielolokalnych" polskich naukowców, Prace i Studia Geograficzne, 68.2, Wydział Geografii i Studiów Regionalnych Uniwersytetu Warszawskiego, Warszawa, 29–53. https://doi.org/10.48128/pisg/2023-68.2-02

King R (2012) Geography and migration studies: retrospect and prospect. Popul Space Place 18:134–153

King R, Cela E, Fokkema T (2021) New frontiers in international retirement migration. Age Soc 41(6):1205–12020. https://doi.org/10.1017/S0144686X21000179

King S (2002) Toward a new map of European migration. Int J Popul Geogr 8:89–106

Kofman E, Raghuram P (2022) Gender and migration. In: Scholten P (eds) Introduction to migration studies. IMISCOE Research Series. Springer, Cham. https://doi.org/10.1007/978-3-030-92377-8_18

Massey DS, Arango J, Hugo G, Kouaouci A, Pellegrino A, Taylor JE (1994) International migration theory: the North American case. Population and Development Review

McAuliffe M, Triandafyllidou A (eds) (2021) World migration report 2022. International Organization for Migration (IOM), Geneva

Milanovic B (2007) Globalization and inequality. In: Held D, Kaya A (eds) Global inequality: patterns and explanations. Polity, Cambridge and Malden, MA, pp 26–49

Van Mol C, de Valk H (2016) Migration and immigrants in Europe: a historical and demographic perspective. In: Garcés-Mascareñas B, Penninx R (eds) Integration processes and policies in Europe. IMISCOE Research Series. Springer, Cham. https://doi.org/10.1007/978-3-319-21674-4_3

Neidhardt AH, Butcher B (2022) Disinformation on migration: how lies. In: Half-trust and mischaracterizations spread. https://www.migrationpolicy.org/article/disinformation-migration-how-fake-news-spreads (13.09.2022)

OECD (2020) Diversity at work: making the most out of increasingly diverse societies. https://www.oecd.org/els/diversity-at-work-policy-brief-2020.pdf

Olsson P, Galaz V, Boonstra WJ (2014) Sustainability transformations: a resilience perspective. Ecol Soc 19(4):1. https://doi.org/10.5751/ES-06799-190401

Papademetrious DG, Benton M (2016) Towards a whole of society approach to receiving and settling newcomers in Europe. Migration Policy Institute https://www.migrationpolicy.org/sites/default/files/publications/Gulbenkian-FINAL.pdf

Polish Border Guards (2022) Official tweeter account updated daily. https://twitter.com/Straz_Graniczna

Serraglio DA, Ferreira HS, Robinson NA (2019) Climate-induced migration and resilient cities: a new urban agenda for sustainable development. Seqüência Estudos Jurídicos e Políticos 42(83):10–46. https://doi.org/10.5007/2177-7055.2019v41n83p10

Skeldon R (2012) Migration transitions revised: their continued relevance for the development of migration theory. Popul Space Place 18(2):154–166

Sobczak-Szelc K, Fekih N (2020) Migration as one of several adaptation strategies for environmental limitations in Tunisia: evidence from El Faouar. CMS 8:8. https://doi.org/10.1186/s40878-019-0163-1

Stoler J et al (2021) Connecting the dots between climate change, household water insecurity, and migration. Curr Opin Environ Sustain 51:36–41

UN DESA (1998) United Nations, recommendations on statistics of international migration, Revision 1, Statistical Papers, Series M, No. 58, Rev. 1. Sales No. E.98.XVII.14

UN DESA (2021) United Nations, department of economic and social affairs. In: International migration 2020. Highlights UN, New York

UN SD (2017) United Nations department for economic and social affairs statistical division. In: Handbook on measuring international migration through population censuses. Demographic statistics. UN, New York

UNDP (2009) Human development report 2009: overcoming barriers: human mobility and development. United Nations Development Programme, New York

UNHCR (2022) Operational data portal—Ukraine refugee situation. https://data.unhcr.org/en/situations/ukraine?fbclid=IwAR3pZAofxq3XIwjcdIy-jqrQaIe8kCTMeyGDk3I8DY8Jx3JqWjJD6IDCZ10

UNHCR (2022) Protecting forcibly displaced and stateless children: What do we know? UNHCR's child protection data from 2015–2021. https://www.unhcr.org/sites/default/files/2023-05/child-protection-data-analysis-2015-2021.pdf

United Nations High Commissioner for Refugees (UNHCR) (2020) Global trends: forced displacement 2019. Geneva, Switzerland. Available from www.unhcr.org/globaltrends2019

United Nations Relief and Works Agency for Palestine Refugees in the Near East (UNRWA) (2020) UNRWA in figures as of 31 Dec 2019. www.unrwa.org/sites/default/files/content/resources/unrwa_in_figures_2020_eng_v2_final.pf
UNHCR (2015) The year of Europe's refugee crisis. https://www.unhcr.org/news/stories/2015/12/56ec1ebde/2015-year-europes-refugee-crisis.html
UNODC (2018) Global report on trafficking in persons 2018. Sales No. E.19.IV.2. Available from www.unodc.org/documents/data-and-analysis/glotip/2018/GLOTiP_2018_BOOK_web_small.pdf
Van Pottelberge A, Caestecker F, van de Putte B, Lievens J (2021) Partners selection patterns in transition: the case of Turkish and Moroccan minorities in Belgium. Demographic Res 45(34):1041–1080. https://doi.org/10.4054/DemRes.2021.45.34
Weichselbaumer D (2016) Discrimination against female migrants wearing headscarves. IZA DP. No 10217. http://ftp.iza.org/dp10217.pdf
Williams AM, King S, Warnes AM (1997) A place in the sun: international retirement migration from Northern to Southern Europe. Eur Urban Reg Stud 4:115–134
WMR (2019) World migration report 2020. International Organisation for Migration
Yang WS, Chia-Wen LuM (2010) Asian cross-border marriage migration: demographic pattern and social issues. Amsterdam University Press, Amsterdam
Zelinsky W (1971) The hypothesis of the mobility transition. Geogr Re 61:219–249.01

Chapter 3
Selected Economic Issues in the Contemporary World

Katarzyna Podhorodecka and Magda Głodek

3.1 Introduction

In this chapter, the selected economic issues in the Contemporary World were shown. In the first part it was the occurrence of the global economic crisis on pace or growth and the development of the global tourism sector with the regional diversification of the global economic crisis in tourism and with the analysis of the economic impact of COVID-19 pandemic on the pace of global economic growth and the global tourism sector. Second part was devoted to the problem of tax havens around the world especially for well-developed countries with the changes in the distribution and list of tax havens and presentation of typology and socio-economic characteristics of tax havens. The last part of the chapter was shown the problem of selling citizenship by selected countries (mainly microstates).

3.2 Influence of the Global Economic Crisis on Pace or Growth and the Development of the Global Tourism Sector

After the global economic crisis (2008–2010), the world economy was developing very quickly. Many economists were saying that the next economic crisis is coming, but nobody was expected that it will be connected with health threats such as virus COVID-19. The pandemic was a huge problem, especially for the hotels, restaurants, and entertainment sector. The multiplier effect of that sector was also visible with bed

K. Podhorodecka (✉) · M. Głodek
Faculty of Geography and Regional Studies, University of Warsaw, ul. Krakowskie Przedmieście 30, 00-927 Warszawa, Poland
e-mail: kpodhorodecka@uw.edu.pl

K. Podhorodecka and T. Wites (eds.), *Global Challenges*, Springer Geography,
https://doi.org/10.1007/978-3-031-60238-2_3

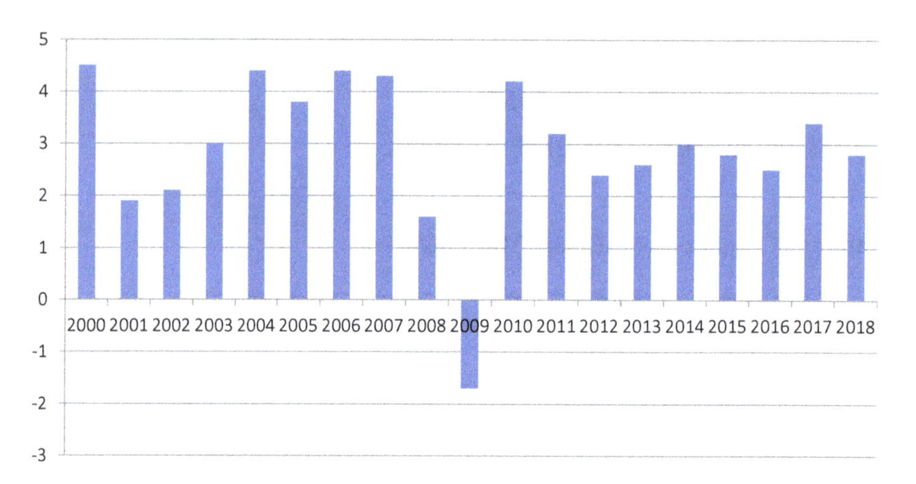

Fig. 3.1 Changes of the Gross National Product for all world economies in the years 2000–2018 (in %). *Source* Own elaboration on basis of data of World Bank—www.data.worldbank.org

economic results. Figure 3.1 is presented the changes of the Gross National Product for all world economies in the years 2000–2018.

The global economic crisis (2008–2010) strongly affected the volume of tourism in the world. This section presents the impact of the global economic crisis on the tourism sector in the world. It highlights which tourist destinations suffered the most during the global economic crisis and which areas did not emerge for years from the recession associated with the global economic crisis. In addition, the conditions affecting the reaction of the tourist sector to the global economic crisis are also discussed, as are the dependent countries and territories that, paradoxically, recorded an upward trend in incoming tourism during the crisis.

In the last 30 years, before the 2020—COVID-19, none of the epidemiological, natural, or terrorist events has had such a big impact on the tourism sector as the global host economic crisis. Even the attacks on the World Trade Center in 2001 in the USA did not have such a strong and long-lasting impact on the global tourist market. Economic crises, like political ones, belong to the category of crises caused by human activity. Their impact on the tourism sector has been widely discussed. According to Okumus and Karamustafa (2005), political crises affecting tourism include political unrest in Fiji and Tibet as well as the war in Iraq in 2003. The political crisis connected with the annexation of Crimea by Russia in 2014 had a bearing on the tourism sector in this region. Stylidis and Terzidou (2014) pointed out that in times of crisis, the problem is uncertainty about jobs. It is important to jointly plan and create appropriate strategic documents for the needs of residents and the tourism industry. This approach results in the support of the local population for the development of the tourism sector (Stylidis and Terzidou 2014). According to Eugenio-Martin and Campos-Soria (2014), households making domestic trips may not react to the economic crisis, but only try to reduce tourist expenses in the place visited. Bronner and De Hoog (2012) pointed out that, during the crisis, effective strategies for dealing with it and having relevant information were important. In the literature, it was also pointed out that during the crisis, enterprises with flexible specializations and diversified products functioned better (Dahles and Susilowati 2015).

This chapter indicates, which tourist destinations suffered the most during the global economic crisis. It began in the USA and was felt in almost all countries around the world. It reached Europe almost a year later. The source of the crisis was in the real estate market in the USA, then the banking sector was affected, followed by the automotive sector, and the crisis was noticeable in almost every area of the economy (Rosati 2009; Zdon-Korzeniowska and Rachwał 2011). Along with the deterioration of the situation in some households, the crisis was noticeable by the significant reduction in demand. In a situation of uncertainty, people start to save and behave more rationally in terms of spending money. In the case of tourist services, they began to limit the number of trips during the year, choose closer tourist destinations, often giving up foreign travel for domestic travel. The need to save money also affected the choice of accommodation. Tourists chose accommodation with their relatives and friends or accommodation at campsites instead of in a hotel (Dziedzic et al. 2010). Globally, the number of foreign trips in 2009 decreased by 4% in comparison to 2008 (from 920 to 880 million) (UNWTO World Tourism Barometer, 8(1), 2010) and tourist expenses by 8% (UNWTO 2010). At the beginning, it was noticeable in the tourism sector's annual data in 2009. Later, it also resulted in demand slowing in 2010. Some countries did not even emerge from the crisis in 2011–2015 and fell into long-term recession. According to Zdon-Korzeniowska and Rachwał (2011), the total global tourist demand decreased by over 10% in 2009 and amounted to approximately USD 7.1 trillion. According to Kachniewska (2012) due to the effects of the global economic crisis, a large number of tourist entrepreneurs were reducing the cost of operations.

3.2.1 *Regional Diversification of the Global Economic Crisis in Tourism*

The regional diversification of the global economic crisis in tourism was very large, and the situation was more complicated when individual countries and dependent territories were taken into consideration. The only region that did not feel the effects of the global economic crisis in the tourism sector in 2009 was Africa, which reacted with an increase in tourist movement of 5% compared to 2008. However, the crisis was very visible in the tourist market in Europe—and almost 6% decline in tourist movement in North America—a decrease of 5%, in the Asia–Pacific region—a decrease of over 2%, and in the Middle East—a fall of almost 6% (Dziedzic et al. 2010; UNWTO World Tourism Barometer 2010). The European market was responsible for 70% of the decrease in the number of international trips in the world. Table 3.1 presents the ranking of countries with the largest increases and decreases in the number of arrivals of foreign tourists in 2008–2009. The dependent countries and territories with the largest declines in the number of foreign tourist arrivals were Madagascar, Bangladesh, Saint Kitts and Nevis, Guatemala and Brunei, with the highest increases in the number of tourist arrivals found in Togo, the Central African Republic, and Micronesia.

Figure 3.2 shows the number of arrivals of foreign tourists and the volume of expenses of foreign tourists in the world.

It can be observed that while the fall in the number of tourists had already begun to rise in 2010, the rise in the level of expenses for foreign tourists in the world before

Table 3.1 Ranking of countries with the largest increases and decreases in the number of foreign tourist arrivals in 2008/2009

Ranking of the biggest decreases (the first 10 countries with the largest decline)	Ranking of the largest increases (the first 10 countries with the largest increase)
– Madagascar—decreased by 57% – Bangladesh—decreased by 43% – Saint Kitts and Nevis—decreased by 35% – Guatemala—decreased by 31% – Brunei Darussalam—decreased by 31% – South Africa Republic—decreased by 27% – Saudi Arabia—decreased by 26% – Comoros—decreased by 24% – Latvia—decreased by 21% – El Salvador—decreased by 21%	– Togo—increased by 103% – Central African Republic—increased by 71% – Micronesia—increased by 85% – Congo—increased by 52% – Haiti—increased by 50% – Lebanon—increased by 38% – Albania—increased by 37% – San Marino—increased by 31% – Angola—increased by 24% – Mozambique—increased by 22%

Source UNWTO database (www.unwto.org)

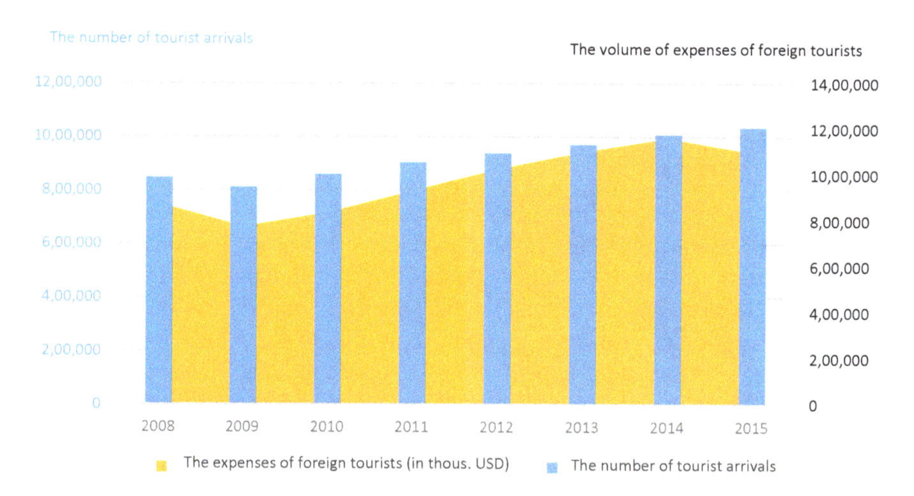

Fig. 3.2 Number of arrivals of foreign tourists and the volume of expenses for foreign tourists in the world in 2008–2015 (in thousands of USD). *Source* Iwańczak B, Podhorodecka K based on www.unwto.org

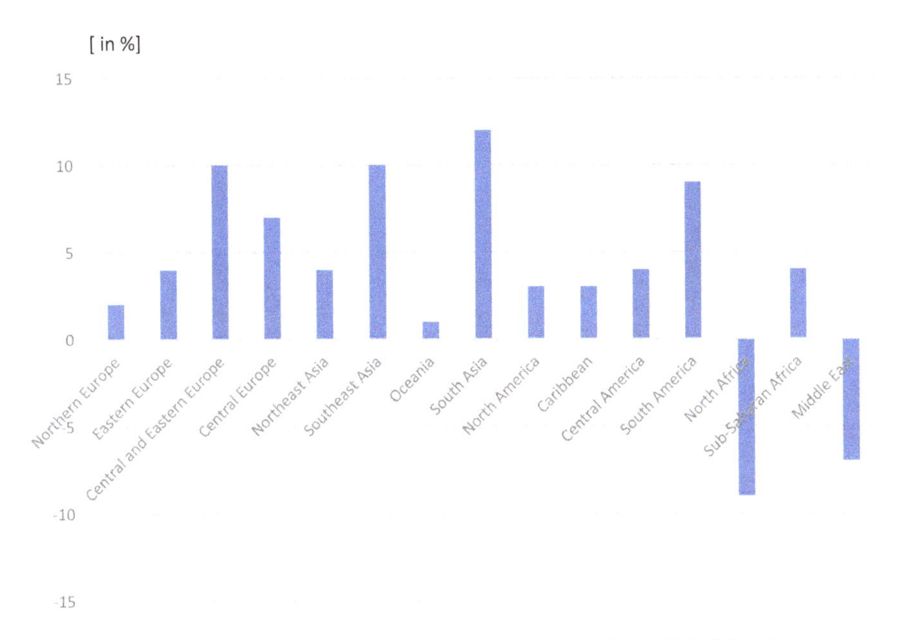

Fig. 3.3 Change in the number of foreign trips in the world by region in 2011/2010 (in %). *Source* Based on UNWTO data www.unwto.org

the global economic crisis was achieved only in 2011. Figure 3.3 presents changes in the number of international travels in the world by region in 2011 compared to 2010.

It is possible to indicate regions that quickly started to gain tourists, thereby compensating for the global economic crisis. Out of 164 countries and dependent territories for which UNWTO collects data, as many as 107 registered a decline in the number of tourist arrivals in 2009 compared to 2008. A drop below 0.2% was registered for 32 countries. Stagnation at the level of 0–0.1% was registered for 20 dependent countries and territories. On the other hand, there was an increase of over 3% for 35 countries only.

Figure 3.4 presents the changes in the number of arrivals of foreign tourists in the world 2009/2008. It clearly shows the area of the African continent which, in the beginning, does not actually recorded the crisis.

Figure 3.5 presents changes in the volume of expenses for foreign tourists in the world in 2009 compared to 2008.

There are definitely a few countries of South America and Africa here. Out of 222 countries and dependent territories for which UNWTO collects data, data on tourism expenditure in 2008–2009 was available for 148. Increases in tourism expenditure were registered for 40 countries, including increases above 20% for 17 countries. On the other hand, declines in tourist expenses were recorded for 107 countries and dependent territories, including decreases below 20% for 22 countries.

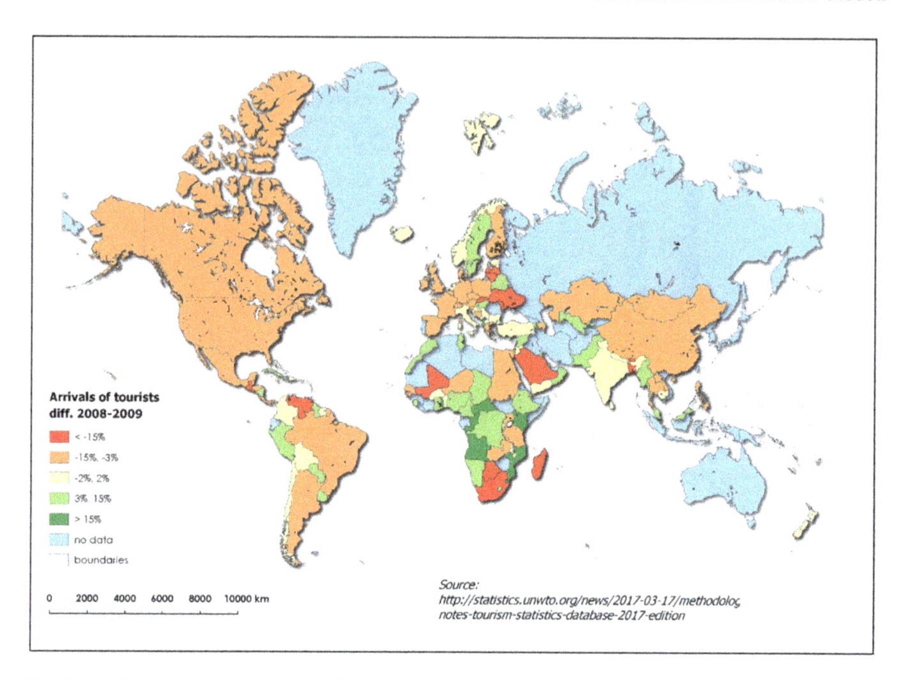

Fig. 3.4 Changes in the number of arrivals of foreign tourists in the world 2009/2008. *Source* Iwańczak B, Podhorodecka K based on UNWTO data (www.statistics.unwto.org)

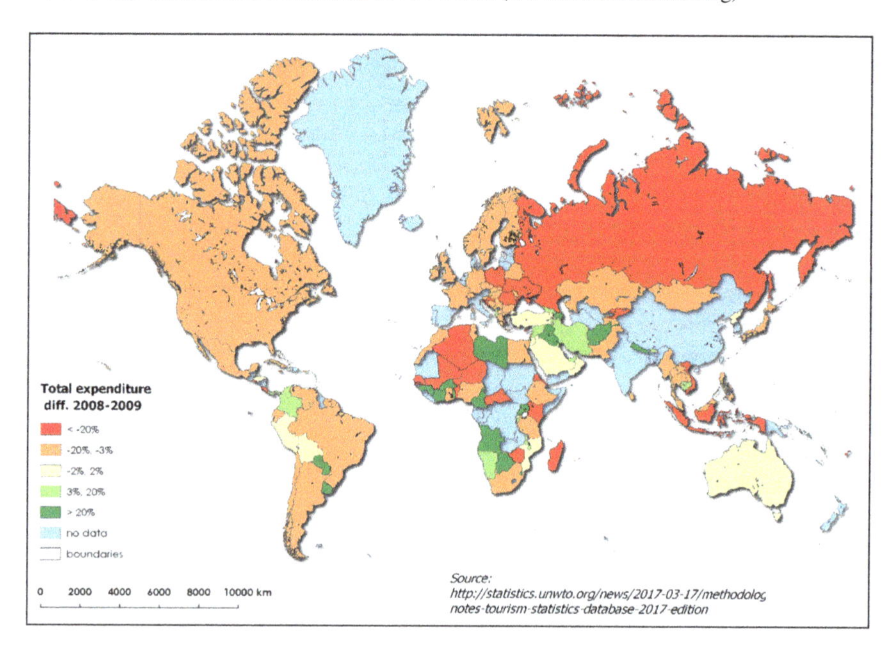

Fig. 3.5 Change in the expenditures of foreign tourists in the world 2009/2008 in USD (in %). *Source* Iwańczak B, Podhorodecka K based on UNWTO data (www.statistics.unwto.org)

Although the crisis was noticeable in tourism, in most countries and dependent territories in the world, one can indicate the areas that emerged from it successfully. At the very beginning, the global economic crisis in the tourism sector did not reach Africa. There were also countries that did not react at first to the decline in tourism including, for example, Togo, Central Africa, and Micronesia. For over 100 countries and dependent territories, the number of tourist arrivals in 2009 decreased in comparison to 2008. Stagnation in tourist movement was registered for 20 countries and dependent territories, while increases above 30% were registered only for 8 countries and dependent territories. The areas with the largest decreases in the number of arrivals of foreign tourists were Madagascar, Bangladesh, and Saint Kitts and Nevis.

3.3 Economic Impact of COVID-19 Pandemic on the Pace of Global Economic Growth and the Global Tourism Sector

The economic impact of COVID-19 pandemic in 2020 was enormous all over the world. Not only well-developed countries were affected but also developing countries due to the fact of existence of global economic crisis. The biggest losses were recorded in the tourism and entertainment sector. Prognosis of the GDP growth rate is very low for many countries. The second wave of COVID-19 results in further verifying economic prognosis. Many economists have stopped believing in the scenario the letter V—that is, a one-time deep decline and a quick rebound, and now the U-shaped—in which GDP stays down for longer (www.businessinsider.com.pl). The third wave of COVID-19 pandemic seems to be enormous except or countries with high level of vaccinates against COVID-19 virus (i.e., Israel, the USA). Škare et al. (2020) claim that pandemic crises result in long negative impact on economy and specially tourism industry. According to Sigala (2020), COVID-19 pandemic results in many socio-cultural, economic impacts on tourism industries (Sigala 2020). Decisions about quarantine and closing the borders of selected counties were bringing the huge loses for tourism industry (Altunta and Gok 2020).

According to European Commission and prognosis from November 2020, the Euro countries' economy will decrease by 7.8% in 2020 and in 2021 will increase by 4.2% (in 2022 by 3%). In turn, the European Union economy will decrease in 2020 by 7.4% and increase in 2021 by 4.1% (in 2022 by 3%). Job losses and rising unemployment are severely limiting the living standards of many Europeans. Unemployment rate in the euro area countries is predicted to increase from 7.5% in 2019 to 8.3% in 2020 (9.4% in 2021, 8.9% in 2022). In the European Union countries, the unemployment rate is expected to increase from 6.7% to in 2019 to 7.7% in 2020 (8.6% in 2021, 8% in 2022) (www.ec.europa.eu/info). Figure 3.6 shows the forecasts about the GDP growth and the unemployment rate in EU countries in 2020–2022.

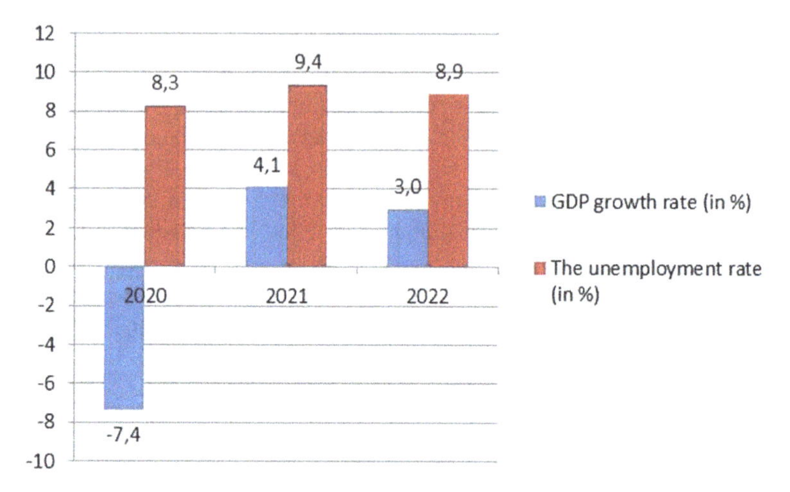

Fig. 3.6 Forecasts about the GDP growth and the unemployment rate in EU countries in 2020–2022 (in %). *Source* Own elaboration on basis of www.ec.europa.eu/info

According to the European Commission—*A Coordinated Economic Response to the COVID-19 Epidemic*, the scale of the impact of COVID-19 pandemic on the EU tourism industry is enormous. Europe is dealing with a significant drop in the number of arrivals from outside the EU (mass cancelations of the travels by tourists from the USA, China, Japan, and South Korea). There was also recorded a huge decline in tourism among EU countries. The small enterprises sector will be mostly affected in tourism economy especially because of decline of business travel. The situation is changing by the disruption of travels inside the European Union and domestic travel (accounting for 87% of arrivals) since March 2020. The most severe is affected the MICE sector—more than 220 huge events in Europe were cancelled or postponed. The COVID-19 pandemic and efforts also result in the decline on other related sectors: food and beverage and cultural activities (www.eur-lex.europa.eu/legal-content).

According to Forbes, COVID-19 spreads more easily in economy than among humans (especially, the first wave). The global downturn may be even greater than the economics said in the beginning (www.forbes.pl). Figure 3.7 is presented the pace of the GDP growth in the world in the years 2000–2019. The huge drop was recorded in 2009 as an effect of global economic crisis.

What are the outlooks for the economy after the COVID-19 pandemic. In the European Union countries is prepared 'Recovery and Resilience Facility' (RRF). The Recovery and Resilience Facility will bring 672.5 billion euro in loans and grants available to support reforms and investments undertaken by countries of the European Union. The aim of this plan is to mitigate the economic impact of COVID-19 pandemic and to prepare European economies to be more sustainable and resilient (www.ec.europa.eu/info). The RRF will also help countries of EU to effectively take

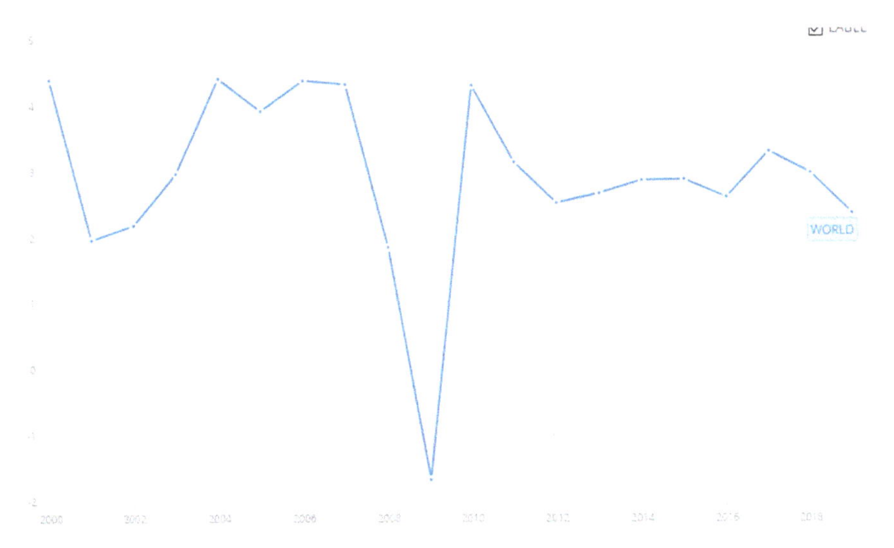

Fig. 3.7 Pace of the GDP growth rate in the world in the years 2000–2020 (in %). *Source* World Bank—www.data.worldbank.org/indicator

the challenges identified in countries and to coordinate the economic and social policy (www.clustercollaboration.eu).

RRF is a comprehensive document that defines goals related to the reconstruction and creation of socio-economic resilience in Poland after the crisis caused by the COVID-19 pandemic. It formulates proposals for reforms and investments that will help in returning to the proper functioning of the Polish economy. The RRF the basis for the use of the Facility for Reconstruction and Increasing Resilience is worth EUR 750 billion at the level of all EU member states. From this pool, Poland will have at its disposal approximately EUR 58 billion, including EUR 23.9 billion in subsidies and EUR 34.2 billion in loans. Funds under the RRF will be invested in five thematic areas: resilience and competitiveness of the economy—PLN 18.6 billion, green energy and reduction of energy consumption—PLN 28.6 billion, digital transformation—PLN 13.7 billion, availability and quality of health care—PLN 19.2 billion, green and intelligent mobility—PLN 27.4 billion. As part of the RRF, there will be opportunities for obtaining funding: for the tourism industry in component: 1: resilience and competitiveness of the economy—support for investments in enterprises aimed at changing/expanding the business profile of entities from sectors most affected by the COVID-19 pandemic. The anticipated amount of the allocation is EUR 300 million (www.gov.pl/web/rozwoj-praca-technologia).

3.4 Global Problem of Tax Havens for Developing Countries

This chapter analyzes the problem of tax havens in the world and the geographical conditions affecting individual countries and dependent territories that implement them. The changes in the number of tax havens in the world have been shown with their geographical location which has changed. This chapter highlights how to counteract the creation of tax havens in the world, from what countries people decide to use the function of a tax paradise and what prompts them to make such a decision. The number of tax havens is shrinking formally in the world; out of their total number, the share of countries belonging to the European Union is declining. People who hide their assets in tax havens come mainly from countries with high GDP per capita, and the size of hidden deposits in these areas is constantly growing. Only legal regulations and efficient fiscal control may contribute to reducing the size of deposits in tax havens. The possibility of depositing funds in tax havens is used by entrepreneurs and the legal profession from countries with low fiscal morals and from countries with a high level of taxes. Moreover on the basis of a list of tax havens and selected criteria, a typology of these areas was prepared. Three typological criteria were analyzed: GDP per capita, share of services in GDP, and number of years since independence.[1]

3.4.1 *Occurrence of Tax Havens in the World*

Over the past two decades, many researchers have dealt with the issues of tax havens in the world (Hines and Rice 1994; Errico and Musalem 1999; Masciandaro 1999, Dharmapala and Hines 2009, Hines 2006; Schwarz 2011). The Organization for Economic Cooperation and Development (OECD) has prepared many studies on the harmful impact of these areas on the international economy. These include Harmful Tax Competition: An Emerging Global Issue (1998), Towards Global Tax Cooperation (2000), and The OECD Project on Harmful Tax Practices (2004).

According to Starchild (1944), a tax haven is an area whose legal system, tradition, and agreements with other foreign countries result in the reduction or elimination of financial burdens (Sikorski 2017; Mazur 2012; Starchild 1993). According to Palan et al. (2010), it is important to look for reasons for the 'commercialization of independence' of individual areas and the adoption of tax haven functions. The conditions for creating tax havens are not only the amount of taxes, but also the relevant legislation and the rapidly integrating international market.

This chapter describes the changes in the number of tax havens and their geographic location in the world. Efforts were made to identify how the creation

[1] This chapter was prepared on the basis of master thesis defended in University of Warsaw, Faculty of Geography and Regional Studies in 2019 by Magda Głodek (Maiden name: Białek)—Typology of tax havens in the world (Białek 2019).

of tax havens can be counteracted and from what countries people decide to use the tax haven functions and what causes them to make such a decision. The possibilities of finding ways to counteract the creation of tax havens were also discussed. The number of tax havens is shrinking formally in the world. The share among EU territories is decreasing, while the size of deposits is constantly growing. Only legal regulations and efficient fiscal control may contribute to reducing the size of deposits in tax havens. Mainly companies and individuals from countries with low fiscal morals and high taxes enjoy the function of tax havens.

According to Mara (2015), low taxation is not the only factor affecting the creation of a tax paradise. A large share of services in the creation of GDP is also a prerequisite for creating good conditions in this field. Developing the function of tax havens for many countries, and micro-countries in particular, is a great economic opportunity. For other highly developed countries, however, it is a threat, because some taxes will not affect their budgets and will be directed to tax havens.

The concept of tax haven is not clearly defined and uniformly defined. The exact definition of this concept is very complicated, mainly due to the lack of a clear and unambiguous framework in international law. In English, the following terms are used: 'tax haven', 'offshore financial center', 'offshore tax haven', in French 'paradis fiscal', in German 'steuerparadiese', and Italian 'paradisi fiscali'. However, the most frequently used term is 'tax haven', which in English means shelter, harbor (Lipowski 2002; Nawrot 2011). However, the concept of offshore financial center (OFC) has a broader meaning range from tax haven, because in addition to tax issues, it includes other important factors that influence the choice of tax paradise (Lipowski 2002).

Refugees, asylum, or tax oases are commonly used in relation to a group of countries or territories that create particularly favorable conditions for foreign investors to invest (Kurcil-Białecka 2011). The beginnings of tax havens date back as far back as the sixth-century BC, when the area was first recognized as the island of Rhodes, then Delos. The territories that offered tax benefits also existed in the Middle Ages—in Hanseatic cities, including coastal duty zones and markets exempt from trade taxation. Tax havens gained more importance during the French Revolution, when wealthy people deposited their money in countries with lower taxes. In the seventeenth century, English colonies in North America served as paradise. Even leaflets were even published there, which promoted these countries as an alternative to Europe, in which strict tax rules were in place. However, the real bloom of tax havens occurred only in the second half of the twentieth century (Kuchciak 2012; Byrska and Borowska 2017).

The Organization for Economic Cooperation and Development (OECD) has created a commonly used definition of a tax paradise, which thanks to its authority is widely accepted. The 1998 OECD report defining the notion of a tax haven states that a tax haven should be a state and a tax-independent territory that applies taxes or other non-tax instruments to trigger economic activity in the service and financial sector. This is quite a general definition that needs to be expanded for returns contained in specialized publications.

According to Lipowski (2004), tax havens are states or dependent territories that apply certain fiscal privileges in order to increase the inflow of foreign capital. An interesting definition was also developed by Starchild, who considers a tax haven to be a country whose legal system, tradition, regulations, and even agreements with other foreign entities make it possible to reduce financial burdens, even to a full extent (Sikorski 2017). This is mainly due to the desire to attract foreign capital, which immediately attracts the interest of banks, insurance companies, and trusts to these areas and, as a consequence, contributes to the dynamic development of the economy and the creation of new jobs (Kuchciak 2007).

In the narrow sense of the definition of a paradise, it is an area that primarily offers lower tax rates compared to other countries. In the broad sense, a tax territory is considered to have:

- zero or very low taxation;
- liberal local laws when it comes to registering and running a business;
- economic and political stability;
- agreements to avoid double taxation;
- a network of well-functioning companies that provide services in the fields of law, consulting and accounting;
- legal guarantees in the event of nationalization and expropriation;
- laws that recognize the status of a foreigner;
- low costs for buying and maintaining property;
- a favorable location and very well-developed transport infrastructure, banks, and telecommunications;
- strictly protected bank secrecy and business information;
- no currency restrictions, exemption from foreign exchange control (Kuchciak 2007).

According to classical definitions, a tax haven is a country or territory with very low or no tax. However, as Mara (2015) notes that tax havens are not only low or no tax, they are also characterized by a high level of secrecy and a developed network of financial services. There is no statutory definition of a tax paradise in Polish legislation. The legislator only uses the term country or territory using harmful tax competition. The term tax haven functions in the OECD acquis. The provisions of Polish, European, and other laws use the notion of harmful tax competition. However, the meanings of both terms are largely convergent (Nawrot 2011). The concept of tax havens in Polish literature was introduced by Głuchowski in his key work, 'Tax Oases', in 1996 (Głuchowski 1996).

It is worth noting that tax havens are not limited to tax assessment only. Importantly, economic factors do not exhaust the qualities a tax haven should meet. Important criteria for recognizing a given territory as a paradise should also include a favorable geographical location, efficient transport connections, highly developed infrastructure, qualified staff, or professional advisory services (Nawrot 2011). For example, a country with low tax rates will be unattractive if bank secrecy is not sufficiently protected and registration of companies is limited. It can therefore be concluded that an inherent characteristic of a tax haven is a specific investment climate, which encourages capital to be invested in it (Lipowski 2004).

It should also be noted that the concept of a tax haven, which had negative associations traditionally, is losing its negative tone today through progressive globalization. The effect of this is the widespread use of tax havens. The existence of tax havens is sometimes accepted by highly developed countries, who find in them a stimulus that contributes to the development of the economy in a given area (Lipowski 2002). The development of tax havens is connected to decolonization processes, the desire to get rich faster by the newly created states, internationalization of the economy, and globalization processes (Artemiuk 2013).

There are some estimates about the amount that is deposited in tax havens. The value of money hidden there, estimated at the end of the 1960s, amounted to 11 billion USD. At the end of the 1970s, this amounted to approximately 358 billion USD. At the end of the 1980s, only around 400 billion USD was concealed in the Caribbean. By the end of the 1990s, it was over 6 billion USD (Hampton and Christensen 2002). In 2012, the amount hidden in tax havens was estimated at 20–35 trillion USD, which accounted for about one third of global GDP (Preuss 2012). This means that the rate of increase in deposits in banks located in tax havens is huge (www.wiadomosci.gazeta.pl). Figure 3.8 shows the estimated amount deposited in tax havens around the world.

It is estimated that losses due to the functioning of tax havens amount to 50 billion USD per year, which is equivalent to the annual development assistance of highly developed countries to developing countries (Preuss 2012).

The term 'offshore tax haven' refers to companies that operate outside the tax haven and are registered there. It is possible to imagine companies that were trading in clothing in EU countries, paying taxes in Cyprus and applying aggressive tax

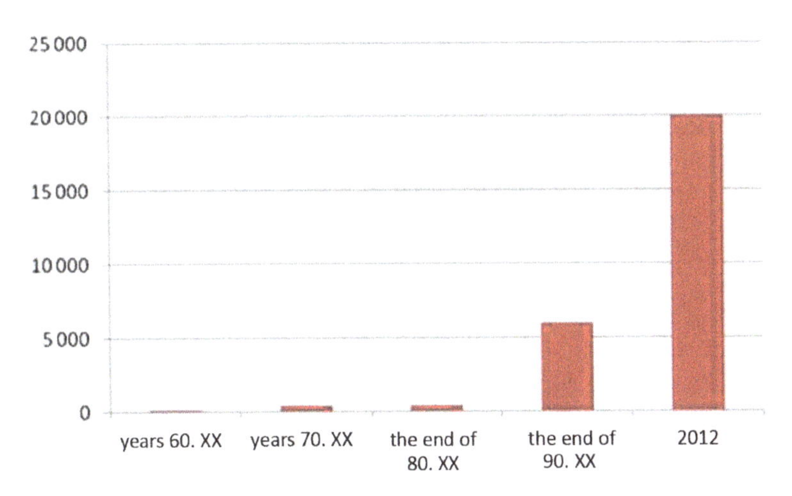

Fig. 3.8 Estimated amount deposited in tax havens in the world (in billions of USD) from the 1960s up to 2012. *Source* Prepared on the basis of Hampton and Christensen (2002); Preuss (2012). www.wiadomosci.gazeta.pl

optimization. Areas that have decided to act as tax havens encourage investors to locate portfolio investments in their territory. As a result, new jobs are created and capital flows into these areas. Even if the taxes are very low, the companies decide to register so much money that the source of income becomes important for selected microstates.

3.4.2 Changes in the Distribution of Tax Havens Around the World

There is a lot of competition in implementing tax havens. In order for individuals to decide to transfer their savings to the chosen location, it must be stable, have permanent financial legislation, and guarantee respect of bank secrecy. It is also important to quickly register a company in a given area (Mara 2015). In many territories, any breach of bank secrecy is treated as a crime (Mara 2015). Building trust among clients can take many years (Lipowski 2002). Many microcities combine several functions by diversifying their economies, i.e., tax havens, tourist havens, earning on the sale of citizenship, sales of votes in international organizations (Bertram 2004; Jędrusik 2005).

Tax havens attract foreign investors, in particular portfolio investments, creating favorable conditions for investing capital. Thanks to this, jobs are created in the areas concerned of these countries and in dependent territories. Each region of the world has its own tax havens because of its geographical location. For North America these are in the Caribbean; for Australia and Japan, the Pacific Islands; for Asia and Africa, islands in the Indian Ocean. The most popular tax havens are the Caribbean—the Cayman Islands, the British Virgin Islands and the Bahamas; the Pacific—Vanuatu and the Cook Islands; and Mauritius and the Seychelles in the Indian Ocean.

In 2016, the scandal known as 'the Panama Papers' broke out, showing the role of tax and financial centers in the global economy (Balakina et al. 2017). This resulted in increased interest in banking secrecy and in the international network that hides its illegal and even derived income (Balakina et al. 2017). In developed countries, the need to create a list of territories that use harmful tax competition has long existed. Panama is considered by many journalists as 'the centre of money laundering of the Chicago Mafia' and the place where many international companies hide money (Mara 2015). Another case from 2017, related to tax havens, was named 'the Paradise Papers'. Many documents from tax havens were revealed at that time thanks to the involvement of a large number of journalists. There were often articles in the press about wealthy people, who were greatly outraged when they heard about tax fraud, yet they used the opportunities offered by tax havens themselves. The whole system is created 'for billionaires and their corporations' (Kowanda 2017).

The tax havens appreciated by investors are characterized by high GDP per capita; they are well governed, and they are small states or territories with a small population and with a large share of services in generating gross domestic product (Mara 2015).

Table 3.2 presents a list of dependent countries and territories using harmful tax competition in their tax systems developed by the International Monetary Fund in 2008.

Table 3.3 presents a list of tax havens in the world prepared on the basis of the Regulation of the Minister of Development on May 17, 2017, on determining countries and territories applying harmful tax competition in the area of corporate income tax. There is a significant decrease in the number of dependent countries and territories that are tax havens.

The number of tax havens has decreased significantly over the last 10 years. According to the International Monetary Fund (2008), there were 46 confirmed tax havens in the world (Mara, 2015), while on the 2015 Polish list, there were 31 areas, and on the list from 2017, there were only 27 areas. Although a total of 19 areas were missing from the list, 9 new areas appeared. You can find the areas that were included on the Ministry of Finance's list and were not on the list of the International

Table 3.2 List of tax havens in the world prepared on the basis of data from the International Monetary Fund (2008)

Europe	North America and the Caribbean region	South America	Africa and the Indian Ocean	Asia	Oceania
Andorra	Anguilla	Panama	Seychelles	Hong Kong	Cook Islands
Monaco	Antigua and Barbuda	Belize	Mauritius	Bahrain	Marshall Islands
Liechtenstein	Barbados	Costa Rica		Singapore	Samoa
Gibraltar	St Lucia			Macau	Vanuatu
Isle of Man	Dominica			Lebanon	Nauru
Malta	St Vincent and the Grenadines			Labuan, Malaysia	Niue
Cyprus	Bahamas				Palau
Ireland	British Virgin Islands				
Jersey	Grenada				
Luxembourg	Aruba				
Montserrat	Turks and Caicos				
Switzerland	Bermuda				
Guernsey	Cayman Islands				
	Netherlands Antilles				
13	14	3	2	6	7

Source Mara (2015), Determinants of tax havens. Procedia Economics and Finance

Table 3.3 List of tax havens in the world prepared on the basis of the Regulation of the Minister of Development on May 17, 2017, on determining countries and territories applying harmful tax competition in the area of corporate income tax (2017)

Europe	North America and the Caribbean region	South America	Africa and the Indian Ocean	Asia	Oceania
Andorra	Anguilla	Panama	Seychelles Republic of Liberia	Hong Kong	Tonga
Monaco	Antigua and Barbuda	Curaçao*	Liberia	Bahrain	Cook Islands
Sark	St Lucia		Mauritius	Macau	Marshall Islands
	Dominica		Maldives		Samoa
	British Virgin Islands				Vanuatu
	American Virgin Islands				Nauru
	Grenada				Niue
	Saint Maarten				
3	8	2	4	3	7

* Curaçao as a Caribbean area lying close to South America (about 60 km from the northern shores of Venezuela) was assigned in the table to South America. Source: prepared on the basis of the Regulation of the Minister of Development of 17/05/2017 on specifying the countries and territories applying harmful tax competition in the scope of corporate income tax. *[Rozporządzenie Ministra Rozwoju z dnia 17.05.2017 r. w sprawie określenia krajów i terytoriów stosujących szkodliwą konkurencję podatkową w zakresie podatku dochodowego od osób prawnych]*

Monetary Fund from 2008: Sark, St Kitts and Nevis, American Virgin Islands, Sint Maarten, Curaçao, Liberia, Macau, Maldives, and Tonga.

The literature also cites positive aspects regarding the functioning of tax havens. The administrations of highly developed countries have a lower propensity to raise taxes because of the fear of companies moving to tax havens. This would result in the loss of tax revenues from companies or private individuals that would move from a high-tax area to a tax haven (Preuss 2012; Hufbauer 1992).

3.4.3 A Blacklist of Tax Havens and the Possibility of Preventing Their Creation

The problem is that individual countries do not have legal instruments to counteract and fight the banking secrecy of other countries, except to publish a list of these countries. The aim of the 'blacklist' is to make it easier to see individuals and companies that register activities in a tax haven. When data is published, it causes shame; customers stop buying the company's products because the company does not pay taxes in their own country. In the case of sports or movie stars, the publication of such information may be the beginning of a decline in popularity or a loss of reputation, resulting in lower profits in the future. A negative image can cause many legal and financial problems.

However, for now, a tool such as blacklisting publications can be considered ineffective, as the value of assets deposited in tax havens is constantly growing. The European Union has been trying to create a law against tax havens for many years, especially those within EU countries. A good example is Cyprus which, after the global economic crisis and thanks to EU actions, began to lose customers as a tax haven. The customers began to move to tax havens that are stable and predictable in terms of taxes (investors fled from Cyprus due to activities consisting of the taxation of deposits above EUR 100.000 in 2013—deposits mainly belonging to foreign investors).

Tax havens are not only a tax problem but also a problem in which money that comes from criminal activity (i.e., the weapons trade, drugs, or corruption) can be concealed. This is especially true for developing countries, where dictators or corrupt politicians can illegally transfer money abroad. In the case of the journalistic investigations, the names of many people in power in developing countries appeared. An important element that contributes to limiting the use of tax havens is the system of tax information exchange and international cooperation. This is regulated in Polish Tax Law in Section VIIa of the Tax Ordinance (Sikorski 2017).

3.4.4 The Methodology of Typology of Selected Tax Havens

Recently, tax havens have become the subject of much international interest. Due to the disappearance of many barriers in the flow of people, capital, goods and services, as well as the transformation of the world into a global village, tax havens are gaining in importance and their use is spreading. They evoke the interest, not only of scientists, business people, or politicians, but also ordinary people (Artemiuk 2013).

A typology of tax havens was developed to answer the following questions:

- where are the tax havens located in the world?
- is there a common feature for all countries/territories that are considered tax havens?
- do selected types of tax havens show regularities in spatial arrangement?
- are there dominant types?

As a result of the research work, a list of the types of tax havens in the world was created. The classification of tax havens can be made on the basis of different criteria, which further justifies the fact that the phenomenon of tax havens is complex and has a relative character. The most popular classification in literature is the distinction of a paradise in a general and specialist sense. A tax haven, which operates on general principles, provides the widest range of financial and tax benefits. In this group, taking into account the criterion of imposed taxes, we should distinguish:

(1) countries or territories not imposing any taxes, the so-called no-tax havens, i.e., Cayman Islands, Nauru, Bermuda;
(2) countries or territories that impose low taxes, the so-called low-tax havens, i.e., Bahamas, Andorra.

Among the countries not imposing any taxes, two groups can be distinguished:

- countries that do not tax income and assets; these countries do not impose direct and indirect taxes on natural and legal persons. The budget of their country is based on income from customs duty, property tax, and various types of fees;
- countries applying the principle of territoriality. It is based on the taxation of income that comes from sources located on its territory. Hence, income from abroad is not taxable. In this group, you can also distinguish the parish which, if they are located in this area but operate outside it, gives companies the status of a tax-exempt entity and those that require companies not to be established and operate outside this area (Kuchciak 2007).

This division seems to be insufficient, because there are countries that impose high taxes, but they offer privileges for specific undertakings or entities, the so-called specialized oases, or special tax havens. They constitute the third category of tax havens (Lipowski 2002, 2004; Kuchciak 2007). These benefits are known as holding and are related to running only administrative tasks or activities that take over shares in other companies.

Tax havens can also be divided into six groups:

- countries where income, capital, and property growth are tax-free, and the law is liberal and flexible—Vanuatu, Cayman Islands, Bahrain;
- countries in which tax is levied according to the rules of territoriality; therefore, the tax is levied only on domestic activities—Liberia, Panama, Haiti;
- low taxation areas—Netherlands Antilles;
- areas that apply tax preferences for selected projects—Gibraltar, Anguilla;
- countries using the holding privilege of carrying out administrative activities or taking over shares in other enterprises—Luxembourg, Liechtenstein;
- territories that treat natural persons preferentially—Bermuda, Cayman, Andorra (Kuchciak 2007).

Based on the tax system of a given country, states and territories may still be distinguished where:

- there is no-tax obligation for residents, or income is not taxed;
- income from individuals is not taxed or taxed very lowly;
- taxation covers income that has been obtained exclusively on its own territory;
- a tax is levied only on income earned abroad;
- in their area, there are specific tax reliefs and preferences, depending on the activity conducted, the existence of holding privilege, or the type of entity.

Tax havens can still be classified based on their geographical location:

- coastal states and territories;
- island states and territories.

Although it is not a legal criterion, it has a significant impact on the choice of tax haven. Most often, entities striving to reduce tax burdens benefit from countries close to them. The dominant paradise tax havens are those located on the islands. However, the least successful among them are located in areas with no access to the sea (Grzywacz 2010).

In summary, the criteria for classifying tax havens are diverse. A more accurate analysis of the classification of tax havens indicates that the commonly used criteria are:

- the range of benefits offered;
- the tax status of business entities established in the paradise;
- the amount of tax imposed;
- the geographical location (Grzywacz 2010).

These are mostly legal factors, but there are also factors unrelated to the law, which does not mean that they are less important for entities interested in tax avoidance.

3.4.5 Features of Tax Havens

Among the entities included in tax havens, we can distinguish countries with different tax systems and different international legal statuses. In addition to independent states (i.e., Bahrain, Liberia, Panama), there are also colonial properties (i.e., Bermuda, Gibraltar, Cayman Islands), autonomous territories (Jersey, Isle of Man, Guernsey), associated countries (Cook Islands, Niue), and even special administrative regions like Macau and Hong Kong. Since a large number of former or current British possessions dominate this group, the English legal system is prevalent. Despite such a large stratification, one can distinguish common and characteristic features of the discussed territories (Foks 2001).

Bearing in mind the existing definitions of tax havens, the following should be distinguished:

- very low or zero income tax rates and no customs burdens;
- lack of cooperation of the administration in the scope of exchange of tax information;
- guaranteed and binding banking secrecy;
- privileged treatment of transferred income to the offshore zone, where it applies, i.e., a lower tax rate;
- non-transparent tax regulations that facilitate the use of various tax privileges (Grzywacz 2010).

The exceptionally liberal banking laws allow for the easy, as well as almost anonymous, opening of bank accounts. On the other hand, the obligation to maintain banking secrecy makes it possible to carry out all financial operations safely and discreetly. Legislation is usually created in such a way that the identity of bank owners and bank account balances cannot be determined. Providing this type of information is tantamount to a violation of the existing banking law (Grzywacz 2010).

Wotava (2000) in his publication 'Tax Offices and Offshore Services' lists a number of common features that are common to tax havens:

- tax havens mostly belonged to British colonies and overseas possessions;
- jurisdiction and legal systems come from very readable English law;
- they are often island countries;
- they are distinguished by the infrastructure level;
- they are located near large markets;
- they pay attention to the preservation of anonymity and discretion;
- there is a liberal, entrepreneurial, and legal climate;
- they are characterized by a high level of banking and lack of foreign exchange restrictions.

In the report published in 1998, the OECD Tax Committee performed certain criteria for determining whether certain member states that apply preferential tax systems are harmful (Jarczok-Guzy 2016). The OECD 1998 report the criteria that allow a given territory to be included as a tax haven listing the following features:

- no, or only nominal, income taxation as a key condition;
- lack of transparent legal regulations;
- there is no effective exchange of system information between national authorities and the OECD;
- the tax system is separated from the domestic economy.

In comparison, the International Monetary Fund lists three basic features that allow it to qualify the territory as a tax paradise:

- a financial system based on foreign assets and liabilities;
- the existence of a significant number of financial institutions that are primarily involved in transactions with non-residents;
- the guarantee of strict banking secrecy, low or no taxation and flexible financial regulations (Kuchciak 2012).

Tax havens are also territories with stable economic and political circumstances. They have qualified staff, including financial advisors, accountants, tax advisors. Another important factor in assessing the stability of a tax paradise is the risk of potential international conflicts and wars. The probability of the occurrence of such events seems to be lower in the case of tax paradises located on the islands than on the continents (Grzywacz 2010). It should be noted that what constitutes a tax oasis for residents of one country is not necessarily beneficial to others. The concept of a tax paradise is therefore a relative term (Rydlak 2017).

Most countries and territories using harmful tax competition distinguish between very liberal and non-transparent legislation in the field of commercial and banking law, as well as conducting business. Examples are: the possibility of instant company registration, or even the lack of it, low requirements for initial capital, fictitious company headquarters, no information obligation to local offices, the possibility of setting up anonymous accounts, no control over bank transfers, no obligation to keep company records and accounts (Foks 2001). Some authors also mention other features, such as living standards, patent protection, and convenient legal regulations in the field for economic entities' rights. This is conditioned by the fact that, as already mentioned, there are no fixed parameters allowing for the unambiguous classification of tax havens (Lipowski 2004).

3.4.6 Tax Havens Around the World—The List

Qualifying a given country as a tax haven may take place under the Act if the state imposes taxes below a certain rate or through the confirmed practice of using the tax legislation in force in a given country in order to avoid paying taxes. By virtue of the Act, tax havens are considered as countries imposing taxes below a certain rate or by listing them in the tax list (Kuchciak 2007).

As there is no common definition of a tax haven, there is also no compliance as to how a given country is recognized as a tax haven. In order to facilitate classification, international institutions, private individuals, and states prepare their own lists. They are usually left open, so they can be modified over time. Some countries are removed from the lists as a result of changing external conditions, internal or intended actions, while others appear on them. In addition, the criteria for classifying countries as tax havens for individual lists differ. As a result, countries considered a tax haven according to the authors of one list are not necessarily considered so by the authors of another. Placement of a given country may have negative consequences—on the one hand, it means applying restrictive anti-state norms on the state, and on the other, reduces its prestige in the international arena (Grzywacz 2010). In many countries, lists of tax havens are created, which may have different characteristics and means of application. There are lists that are valid under the Act, i.e., in Germany, Canada, Great Britain, or lower-level acts, such as in Poland. Lists can also be prepared by the relevant tax authorities for their own needs, i.e., in the USA (Artemiuk 2013).

With the publication of lists of tax havens, international institutions and organizations refrained from criticizing the practice for a long time. Although the harmfulness of such practices was recognized, they were not criticized, mainly because of the desire to maintain correct relations with states and territories having the status of a tax paradise. Under the pressure of the countries that suffered the most losses, international institutions decided to take more decisive steps in countering tax havens. The OECD report 'Harmful Tax Competition: An Emerging Global Issue', known as the 'Report on Harmful Tax Competition', was created in 1998 in order to ensure that budget revenues from taxation of member states would be maintained, which

Table 3.4 Tax havens according to the OECD

Countries and dependent territories	
Andorra	Monaco
Anguilla	Montserrat
Antigua and Barbuda	Nauru
Netherlands Antilles	Niue
Aruba	Panama
Bahamas	Samoa
Bahrain	Seychelles
Barbados	St Kitts and Nevis
Belize	St Lucia
British Virgin Islands	St Vincent and the Grenadines
Dominica	Tonga
Gibraltar	Vanuatu
Grenada	Isle of Man
Guernsey/Sark/Alderney	Cook Islands
Jersey	American Virgin Islands
Liberia	Marshall Islands
Liechtenstein	Turks and Caicos
Maldives	

Source OECD, *Towards Global Tax Cooperation: Progress in Identifying and Eliminating Harmful Tax Practies,* Paryż 2000, s. 17, www.oecd.org/tax/transparency/about-the-global-forum/publications/towards-global-tax-cooperation-progress.pdf. (09.08.2018)

gave rise to deeper discussion in the international arena. It also contributed to the establishment of a list of tax paradises in 2000, as had been announced earlier in the report. Countries classified by the OECD as part of a group of tax havens are presented in Table 5. Initially, more than 40 countries and dependent territories were added to the list, but ultimately there were 35 of them (Grzywacz 2010). Bermuda, Cyprus, Cayman Islands, Malta, Mauritius, and San Marino were not included on the list because they undertook a change in their legislation and agreed to adapt to OECD requirements before the end of 2005 (Foks 2001) (Table 3.4).

The list of non-cooperating tax havens is systematically updated, mainly due to the rapid development of the offshore sector in countries not included in it. Listing involves the use by the OECD member states of the remedies indicated by the Financial Committee and a commitment to the elimination of harmful tax practices. The OECD is considered the unquestionable leader among international organizations dealing with the issues of tax havens, taking into account the scope and comprehensiveness of the publications undertaken (Artemiuk 2013).

There is also a list of tax havens in Poland. For the first time, this type of list was created pursuant to the Regulation of the Minister of Finance of 11 December 2000 on the determination of countries and territories applying harmful tax competition. It is extensive, as it covers over 50 countries and territories. These lists take into account the content of arrangements made in this regard by the OECD. The list includes dependent countries or territories identified as countries using harmful tax competition. The regulation also contains a list of countries applying unfair competition in the area of transactions related to administrative and non-material services as well as services of a financial nature.

It lists three groups of areas where tax systems are considered to be harmful:

- countries and territories, irrespective of the type of transaction (this is the majority, covering 40 countries and dependent territories);
- countries and territories, in which harmful tax competition concerns transactions related to administrative services and other intangible benefits;
- countries and territories, in which harmful tax competition concerns transactions related to services of a financial nature (Kuchciak 2007).

Five lists have so far been created in Poland: dated 11 December 2000; 16 May 2005; 9 April 2013; 23 April 2015; 17 May 2017.

There is also a division of lists into black, gray, and white. The 'blacklists' are territories that are considered by the state as a tax haven, though not a tax paradise. The gray lists, on the other hand, contain territories that can be considered a paradise in exceptional situations. These have only been prepared by some countries (Kuchciak 2012). In total, there are around forty countries and territories in the world that can be described as tax havens. However, not all of them are on the list of countries applying harmful tax competition, developed by the Ministry of Finance (Kurcil-Białecka 2011). Differences in the lists of tax havens are taken from different treatments of various tax benefits by individual countries and institutions that prepare the lists. Therefore, the actual number of tax havens is relative and depends on the accepted criteria for recognizing the area as a tax haven (Kuchciak 2012).

3.4.7 The Location of Tax Havens Around the World in the Twenty-First Century

The largest concentrations of territories recognized as tax havens can be observed between North America and South America. A considerable number of havens are also found on the European continent and to the east of the Australian coast in the Pacific Ocean (Byrska and Borowska 2017).

These countries are beyond the control of Polish law in accordance with the Ordinance of the Minister of Finance of 17 May 2017. By grouping tax havens according to geographical location in 2005, a set was created which includes nine territories from Europe, four from Asia, seven island territories from Australia and Oceania. The largest number of tax havens was located in North and South America—as many as twelve, while the lowest was in Africa.

However, when ranking tax havens according to geographical location in terms of parts of the world, the 2017 list included:

- three tax havens belonging to Europe;
- four to Asia;

- ten to North and South America;
- three tax havens belonging to Africa;
- seven islands to Australia and Oceania.

Since Netherlands Antilles was replaced by five territories in 2010, the list of 2017 includes all areas that already appeared in 2005. Hence, it should be noted that the distribution of tax havens has not changed. Only lists, due to restrictions applied by international organizations, including the OECD, the European Union includes a smaller number of territories.

3.4.8 Socio-economic Characteristics of Tax Havens

Tax havens are usually countries (territories) with a small surface and a small population. They are often former colonies, overseas territories, islands, microstates, or developing countries, which count on the inflow of capital in this manner. Most of them do not have funds, natural resources, or developed industry or agriculture. They depend on deliveries from abroad. The collection of large taxes would therefore be unjustified, as the proceeds to the budget would be relatively small. Therefore, it is better to create the conditions necessary to attract foreign capital. This contributes to the rapid development of the economy through the development of the banking sector and internal infrastructure. Tax havens attract huge foreign investments (Lipowski 2002).

The most common motive for introducing a preferential tax system is the lack of strong currency and a poorly developed industry. The geographical location advantage is a great asset for a tax paradise. The location of the paradise is important because it is the territorial location that determines the type of customer of a given tax paradise and usually comes from countries that are close to the tax paradise (Lipowski 2004).

The outline of a tax paradise can be summarized in several points:

- well governed;
- a small country;
- a low population;
- a high GDP per capita;
- a very large share of services in the GDP structure.

According to a report published by the International Monetary Fund in 2008, GDP per capita in almost all tax havens is higher than the global average (global average excluding tax havens). In general, tax havens have less than one million inhabitants, and they are not members of any international organization (Mara 2015).

Table 3.5 presents the basic data on tax havens. The average number of inhabitants amounted to 805.002 people, which confirms that the area of tax havens does not exceed one million inhabitants. However, the average area is 7.8 km^2.

The largest number of residents was recorded in Hong Kong—over seven million people—a territory in which the population density amounted to nearly 7000 people

Table 3.5 Basic data on tax havens (2017)

Number	Dependent country/ territory	The number of citizens	Surface (km^2)	Density of population (persons/square kilometer)	Date of independence *
1	Liberia	4,731,906	111,370	42.5	1847
2	Panama	4,098,587	75,420	54.3	1903
3	Vanuatu	276,244	12,190	22.7	1980
4	Samoa	196,440	2840	69.2	1962
5	Mauritius	1,264,613	2040	619.9	1968
6	Hong Kong	7,391,700	1105	6689.3	Special administrative region of the People's Republic of China
7	Dominica	73,925	750	98.6	1978
8	Tonga	108,020	750	144.0	1970
9	Bahrain	1,492,584	771	1935.9	1971
10	St Lucia	178,844	620	288.5	1979
11	Andorra	76,965	470	163.8	1278
12	Seychelles	95,843	460	208.4	1976
13	Curaçao	161,014	444	362.6	Dependent territories of the Netherlands
14	Antigua and Barbuda	102,012	440	231.8	1981
15	American Virgin Islands	107,268	350	306.5	The US-dependent territory
16	Grenada	107,825	340	317.1	1974
17	Maldives	436,330	300	1454.4	1965
18	Niue	1611	260	6.2	The territory of New Zealand
19	Cook Islands	17,459	236	74.0	The territory of New Zealand
20	Marshall Islands	53,127	180	295.2	1986
21	British Virgin Islands	31,196	150	208.0	Great Britain's territory
22	Anguilla	14,909	91	163.8	Great Britain's territory
23	Sint Maarten	41,109	34	1209.1	Dependent territories of the Netherlands
24	Macau	622,567	30.3	20,546.8	Special administrative region of the People's Republic of China

(continued)

Table 3.5 (continued)

Number	Dependent country/ territory	The number of citizens	Surface (km^2)	Density of population (persons/square kilometer)	Date of independence *
25	Nauru	13,649	20.0	682.5	1968
26	Sark	620	5.5	112.7	Great Britain's territory
27	Monaco	38,695	2.0	19,347.5	1297

* Political dependence is given. *Source* Own elaboration: The World Bank, www.data.worldbank.org (08.02.2018), United Nations Demographic Yearbook 2017, 2018, Department of Economic and Social Affairs, New York; population density—own calculations

per square km in 2017. The smallest number of people inhabited the island of Sark, while the lowest density of population was recorded on the island of Niue—only 6.2 people per square km. Liberia, with an area of 111.3 km^2, was the largest tax oasis. The smallest tax haven in terms of area was Monaco, whose territory amounted to 2 km^2.

Seventeen of the tax havens are independent territories (the year of independence is given in brackets; National Foreign Countries, Ministry of Foreign Affairs):

- Andorra (1278);
- Monaco (1297) from the Republic of Genoa;
- Bahrain (1971) from Great Britain;
- Maldives (1965) from Great Britain;
- Antigua and Barbuda (1981) from Great Britain;
- Dominica (1978) from Great Britain;
- Grenada (1974) from Great Britain;
- Panama (1903) from Colombia;
- St Lucia (1979) from Great Britain;
- Liberia (1847) from the USA;
- Mauritius (1968) from Great Britain;
- Samoa (1962) from New Zealand;
- Nauru (1968) from Australia;
- Tonga (1970) from Great Britain;
- Vanuatu (1980) from Great Britain and France;
- Marshall Islands (1986) from the USA;
- Seychelles (1976) from Great Britain.

The year of independence varied from 1278 (Andorra) to 1986 (Marshall Islands). The earliest independence was obtained from European countries, followed by a country located in West Africa—Liberia, and Central America—Panama. In contrast, as many as thirteen island territories became independent in the second half of the twentieth century. This is particularly true for the Caribbean and Oceania regions.

The political affiliation of dependent territories is as follows:

- Hong Kong—special administrative region of the People's Republic of China;
- Macau—special administrative region of the People's Republic of China;
- British Virgin Islands, Anguilla—Great Britain;
- Sint Maarten, Curaçao—the Netherlands;
- Niue, Cook Islands—territories of New Zealand that are part of the Commonwealth;
- US Virgin Islands—the USA.

In conclusion, the size of the territory of tax havens and their political status allows it to be stated that traditional tax havens are small states, often called microstates. Some of them are semi-sovereign because they do not run foreign policy independently and do not have their own armed forces. Typical tax havens are territories that are dependent on other countries, such as Curaçao (on the Netherlands) or Sark (on Great Britain). These territories, despite dependence on other countries, have autonomy in tax matters, i.e., Anguilla.

An interesting issue seems to be the link between tax havens and Great Britain. The current overseas territories of Great Britain and former British colonies account for as much as 70% of all surveyed areas. This begs the question—what, then, was British colonialism? Hansen and Kessler (2001) in their article 'The Political Geography of Tax Havens and Tax Hells' examined whether the size of the country influences the taxation model. The article explains the existence and characteristics of tax havens through the interaction of political and geographic factors. It has been proven that in a system of independent democratic states, tax systems depend on the relative size of the geographical jurisdiction. Tax havens represent 12.7% of all countries and territories in the world.

3.4.9 Typology of Tax Havens

Research involving the division of collections into more homogeneous groups has quite a rich tradition in geography. As noted by Runge (2006) in antiquity already, the question of looking for both similarities and differences in the formation of various types of phenomena on the surface of the Earth was an important subject of interest for geographers. Geographical problems included analyses of distribution, location, and spatial differentiation. However, the ordering of humans' surrounding reality is the initial condition for the task of asking about the essence of the phenomena studied. Extracting or combining similar objects into homogeneous groups is an opportunity to create order. In this way, the researcher takes the first step on the road to knowing reality. The main research method, which is also the purpose of the work, is typology. Typology consists in distinguishing one or more types within a given set of elements, as well as comparing individual elements in terms of features, grouping, ordering, and division of these elements according to specific types. The word type has been borrowed from systematics and means a specific set of properties or property that objects should have in order to be considered different from others. The concept of type in geography is very common. For example, in the geography of populations, we have types of age pyramids, migration balances and on farms, types of crops, or types of animal production (Runge 2006).

Typology is a procedure, but also the result of grouping a set of objects in predetermined types. The procedure for obtaining the final result can be done by dividing or grouping. In the first case, all elements of the set are treated as a whole. By means of appropriate procedures, they are divided into parts that are the subsets being researched. Each of them should stand out in terms of the features examined with a high degree of homogeneity. Grouping, on the other hand, is a gradual combination of objects, until the desired set of homogeneous groups is obtained, by means of specific procedures (Runge 2006). It is worth mentioning that typology is next to classification and regionalization, a research method used to create collections. The terms typology and classification are sometimes used interchangeably in many publications, especially those that deal with quantitative methods in geography. In the opinion of Romański, the concept of typology is a peculiar case of classification concepts. The singularity lies in the fact that the typological concepts function as a model (Swianiewicz 1989). The classification must fulfill the obligation of completeness and disconnection. It is a grouping in which all possible classes are considered in the research rigging procedure. However, regionalization should be defined as the procedure and effect of separating regions, where grouped spatial units belonging to the same class are not adjacent to each other (Runge 2006) (Table 3.6).

Table 3.6 Tax havens according to OECD

Countries and dependent territories	
Andorra	Monaco
Anguilla	Montserrat
Antigua i Barbuda	Nauru
Netherlands Antilles	Niue
Aruba	Panama
Bahamas	Samoa
Bahrajn	Seychelles
Barbados	St. Kitts and Nevis
Belize	St. Lucia
British Virgin Islands	St. Vincent and the Grenadines
Dominica	Tonga
Gibraltar	Vanuatu
Grenada	Man Island
Guernsey/Sark/Alderney	Cooks Islands
Jersey	American Virgin Islands
Liberia	Marshall Islands
Liechtenstein	Turks and Caicos
Maldives	

Source OECD, Towards Global Tax Cooperation: Progress in Identifying and Eliminating Harmful Tax Practices, Paryż 2000, s. 17, www.oecd.org/tax/transparency/about-the-global-forum/publications/towards-global-tax-cooperation-progress.pdf. (09.08.2018)

Typology does not explain processes and phenomena, but in order to secure data, it leads to generalizations or verification of the previously adopted theories (Swianiewicz 1989). A typology was carried out at work to determine the dominant types of tax havens. An attempt to characterize them will also be undertaken. Typology includes tax havens, which are characterized by characteristic features for the whole set. To implement the typology, the dependent countries/territories were selected, which were included in the Regulation of the Minister of Finance of Poland of 17 May 2017 on the determination of countries and territories applying harmful tax competition in the field of corporation tax. It was decided to choose the list of the Minister of Finance, because it is the only Polish institution that creates this type of collection. The list takes into account the content of arrangements made in this regard by the OECD. Therefore, it was decided to select this list, which eventually included 27 territories.

The research process can be divided into three stages. The first is choosing the right criteria for tax havens. Finding detailed information for individual parishes turned out to be very difficult. In the end, the following three criteria were selected:

- GDP per capita (in USD);
- share of services in GDP (in %);
- number of years of independence.

The first of the selected criteria was GDP per capita measured in USD dollars. The obtained values were divided into three classes, according to the principle of equal class spread, so that similar values of GDP per capita could be in the same class. The result of assigning the letters of the alphabet is shown in the table.

The second selected criterion was the share of services in GDP in percent, which is in the economy in each of the areas studied. In a few cases, there was no data for 2017; hence, the previous years were used. The key for assigning code symbols was shown in the table below. The last of the selected criteria was the number of years of independence. The typological criterion was divided into classes. It was assessed that the longer the paradise is independent, the more financial independence it has. The second stage of the typological procedure was the collection of relevant data. After ordering the database, a table was obtained containing the following data: territory name, GDP per capita, share of services in GDP, and independence in the number of years. Then, each typological criterion was divided into three classes; they were assigned uppercase letters, which were used to create the codes needed to carry out the typology. The last stage was the development and summary of results. Each of the types created was described and interpreted.

On the basis of selected criteria, a three-letter code was created for each tax haven. The number of codes that could be obtained from the three letters of the alphabet (A, B, C) is 27. Table 3.7 shows the key for assigning values to individual types.

However, some of the codes that might have been found did not occur even once, so fewer codes were received. It can also be noticed that some of the codes are much more common than others. Figure 3.9 shows the number of tax havens for particular types.

As a result of typology, 12 types were obtained:

Table 3.7 Key for assigning values to particular types

Type	GDP per capita (USD)	The share of service in GDP (%)	The independent (the number of years)
AAA	A	A	A
AAB	A	A	B
AAC	A	A	C
ABA	A	B	A
ABC	A	B	C
ABB	A	B	B
ACA	A	C	A
ACB	A	C	B
ACC	A	C	C
BAA	B	A	A
BAB	B	A	B
BAC	B	A	C
BBA	B	B	A
BBB	B	B	B
BBC	B	B	C
BCA	B	C	A
BCB	B	C	B
BCC	B	C	C
CAA	C	A	A
CAB	C	A	B
CAC	C	A	C
CBA	C	B	A
CBB	C	B	B
CBC	C	B	C
CCA	C	C	A
CCB	C	C	B
CCC	C	C	C

Source Own elaboration

- Type AAA—tax haven with high GDP per capita and very high share of services in GDP, independent country;
- AAC type of non-reliant territory, with a high GDP per capita and a very high share of services in GDP;
- ABC type of non-reliant territory with high GDP per capita and high share of services in GDP;
- ACB type—non-dependable paradise, with high GDP per capita and medium about the share of services in GDP;

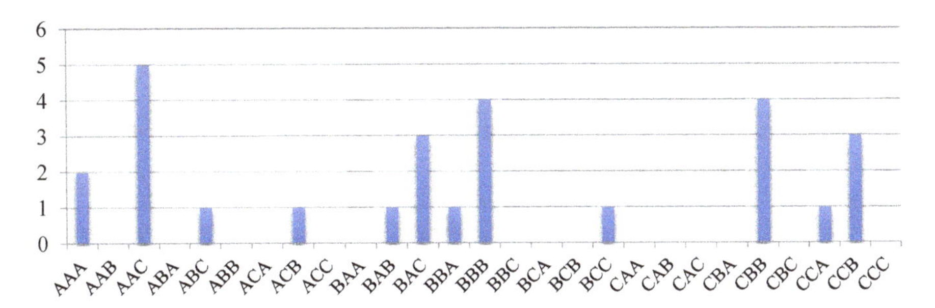

Fig. 3.9 Number of tax havens according to the types received. *Source* Own elaboration

- Type BAB—an independent country less than one hundred years old, with an average GDP per capita and a very high share of services in GDP;
- BAC type—non-reliant territory, with average GDP per capita and a very high share of services in GDP;
- BBA type—an independent country over one hundred years, with an average GDP per capita and a high share of services in GDP;
- Type BBB—an independent country less than one hundred years old, with an average GDP per capita and high share of services in GDP;
- BCC type—non-reimbursing paradise, with average GDP per capita and medium about the share of services in GDP;
- CBB type—an independent country less than one hundred years old, with a low GDP per capita and high share of services in GDP;
- Type of CCA—an independent country over one hundred years old, with a low GDP per capita and the average share of services in GDP;
- Type of CCB—an independent country less than one hundred years old, with a low GDP per capita and the average share of services in GDP.

However, the other types AAB, ABA, ABB, ACA, ACC, BAA, BBC, BCA, BCB, CAA, CAB, CAC, CBA, CBC, as well as CCC did not occur even once.

The most tax havens are represented by the AAC type. They are, therefore, dependent communities, with a high GDP per capita and a very high share of services in GDP. Then, the BBB and CBB types occurred four times, BAB and CCB three times. The AAA type occurred in two cases, the remaining ones after one event.

Analysis of created types —Description of types of tax havens.

Completing a tax paradise to the same type was based on meeting three criteria. For example, type 2 is characterized by:

- a similar value of GDP per capita;
- a similar share of services in GDP;
- a similar number of years of independence, not exceeding 100 years.

Tax havens were classified to the same type if the values of two elements were in line with the symbol. As a result of the typological procedure, the collection of all tax havens was divided into four types.

Type 1 included territories located in Europe—Andorra, Monaco, Sark, two Special Administrative Regions of the People's Republic of China, and two non-autonomous territories—the British Virgin Islands and the American Virgin Islands. Units belonging to this type are characterized by high GDP per 1 inhabitant and a very high share of services in GDP, which exceeds 75%. Andorra and Monaco, one of the smallest independent European countries, have also been included in this type.

Type 2 included almost all areas that gained independence in the second half of the twentieth century, which means exactly when real tax havens flourished. The exception is Panama, which gained independence in 1903. Therefore, the largest group is represented by independent countries with medium or low GDP per capita and very high or high share of services in GDP, over 65%. It is also worth noting that all areas classified as type 2 are independent countries, most of which gained independence from Great Britain.

Type 3 was characteristic primarily for the areas located in the tropic zone. These are the most risky tax havens, lower than other GDP per capita and share of services in the economy < 65%. Types 4 were classified territories, which could not be assigned to the above, because the resulting codes were a combination of the letters ABC. Hence, type 4 is a mixed type.

The presented analysis shows that the highest GDP per capita is found in areas where the services sector plays a decisive role. The analysis shows that the leading countries in terms of GDP per capita are service economies. It is worth noting that financial/banking services are included in the third sector of the economy, hence such a large percentage of this sector in areas that are referred to as tax havens. In addition, the development of the service sector is one from the main trends in modern economies (Węgrzyn 2009). In the case of tax havens (mainly small economies), limited possibilities of economic specialization mean that in most cases, the development of the banking sector seems to be one of the possibilities to break the stagnation socio-economic in this group of territories. For many of them, income from transactions in paradise is often the only source of income. Currently, the services' sector produces the largest part of GDP and employs the highest percentage of employees. Changes in the sectoral structure of employment reflect a continuous change toward a service-based economy and a progressive gradual decline in employment in industry and agriculture. The service sector has been the most dynamic sector in terms of job creation for many years.

3.5 Problem of Selling Citizenship by Selected Countries

Creating income by selling the citizenship is related to the function of tax havens in many island areas. This is related to hiding the assets of wealthy people in areas with very low or no tax rates. People who decide to buy citizenship are very wealthy. They may also try to circumvent their country's sanctions against traveling to highly developed countries. So this means that microstates and small island areas often take profits from selling citizenship. This type of territories is associated with areas that are often full-fledged with decision-making moving their assets to the tax havens, and they are also well-known tourist destinations. In 2016, profits from the sale of citizenship rights to small island towns amounted to 2 billion USD (van Fossen 2017 for CBS 2017). Buying citizenship is also known as acquiring ‘economic citizenship’.

The Caribbean Region is particularly famous for its sales programs. For example, on Saint Kitts and Nevis, since 1984, you can invest a minimum of 250 thousands USD and obtain citizenship. In turn, on Antigua and Barbuda, buying citizenship has been possible from 2013 after investing a minimum of 250 thousands USD. On Dominica, there are no requirements regarding the period of stay on the island, but it is necessary to invest a minimum of 100 thousands USD; their program began in 1993. On Granada, for citizenship, you must make a donation of 150 thousand USD per person or 200 thousand USD for a married couple. It is also possible to purchase appropriate real estate worth 350 thousand USD (www.ceo.com.pl). The island’s citizenship sales program started in 2014 (www.weforum.org, International Monetary Fund). The island territories of the Caribbean region have long been involved in this type of activity. The Saint Kitts and Nevis passport can be purchased by investing in real estate worth at least 400 thousands USD or by supporting a local charity with at least 200 thousands USD (www.polityka.pl).

There is a term ‘shadow nationality’, which means that the person who buys the passport does not have to decide to be loyal to their new homeland, and usually there is no requirement to live in the area (van Fossen 2017; Ronkainen 2011); it is only necessary to make some investments. Unfortunately, it is currently not possible for highly developed countries to monitor the process of the selling of citizenship (van Fossen 2017).

Table 3.8 summarizes the costs and conditions for purchasing citizenship in selected individual island areas and microstates throughout the world.

It is estimated that almost 100 thousands persons obtained the right of permanent residence in one of the countries belonging to the European Union by purchasing citizenship (www.polityka.pl).

In Cyprus, 3 thousand people have purchased passports since 2013, and the island has earned 5 billion euro from the transactions (www.polityka.pl). Many cities in Cyprus, such as Limassol (www.ceo.com.pl), are developing thanks to passport investments.

In Malta, to be able to buy citizenship, it is enough to invest approximately 1.15 million euros, but the great interest in obtaining citizenship meant that the cost was doubled. A one-year residency is also required for a person applying for citizenship of this country (Wójcik 2014). Every year, about 2000 people apply for Maltese passports. If consumers want to acquire citizenship in Cyprus, the minimum amount of investment is at least 300 thousand euro in the real estate market or investment in government bonds with a minimum of 2 million euro. A bank deposit of a minimum of 5 million euro is also possible (www.obserwatorfinansowy.pl). Obtaining citizenship in Malta or Cyprus is an easy way to avoid EU sanctions for residents of Russia.

It is also often said that some of the profits from the sale of citizenship by these countries do not contribute to government funds but are, instead, embezzled by intermediaries (van Fossen 2017).

Table 3.8 Costs and conditions for purchasing citizenship per individual for selected island areas and microstates

Territory	Start date for selling citizenship	Minimum investment to obtain citizenship
Grenada	2014	150 thousands USD for one person, or 250 thousand USD per married couple
Dominica	1993	100 thousand USD
Malta	No data	From 1.15 million euro to 2.3 million euro
Cyprus	No data	300 thousand euro or investment of minimum 2 million euro in government bonds or a bank deposit of at least 5 million euro
Saint Vincent and the Grenadines	No data	The purchase of real estate or aid for charity organization in the sum of at least 400 thousands USD
Saint Kitts and Nevis	1984	250 thousands USD
Antigua and Barbuda	2013	Investment of at least 250 thousand USD

Source Own elaboration on basis: www.polityka.pl, www.ceo.com.pl, www.weforum.org, International Monetary Fund, van Fossen (2017) after CBS 2017

According to *The Guardian*, for poorer countries, functioning as a tax haven is a good solution to getting rid of debts. The International Monetary Fund estimates that St. Kitts and Nevis derive about 14% of its GDP annually from its tax haven function, although there are estimates that indicate that it may be twice as much (www.theguardian.com). Very rich people believe that it has an advantage to have a second, third, or fourth passport. It needs to be mentioned that in island areas and microstates, there are very often several ways to gain profits. This is mainly through the tourism industry, or selling citizenship such as on islands of the Caribbean region.

3.6 Summary and Conclusions

In this chapter, the selected economic issues were characterized. The global of economic crisis on tourism sector was shown (drops in foreign tourists' travel were more than 10% year by year). It contributed serious problems to tourism sector almost for each country and tourism region of the world for several years starting from 2008. Recovery from this type of crisis for international tourism was long. Of course, the impact of COVID-19 pandemic as crisis situation for tourism sector was more huge (drops by 70–80%), but to fully asset it, we need to wait several years to receive the full statistics.

The problem of tax havens is expanding. The estimated size of assets deposited in tax havens is increasing and being a tax haven becomes a way to earn money when there are no other ideas governing more innovative activities. The number of tax havens in Europe has decreased the most, which may be the result of many EU efforts to limit this practice. This is a global problem for all highly developed countries, due to people who earn a lot, who do not want to pay taxes, as well as for developing countries where corrupt politicians place their illegally obtained finances.

In recent years, the list of tax havens has significantly decreased. Analyzing their location, the highest number of tax havens in 2008 was in Europe (13), and in 2017, there were only 3 tax havens left in Asia, and the same number in Oceania. However, new tax havens were created in Africa and the islands of the Indian Ocean. However, the number of tax havens in North America, the Caribbean Region, and South America decreased slightly. The process of globalization of the economy has significantly influenced the relations between tax systems that are in force in individual countries. More and more business entities conduct their activities on an international scale by investing their resources in tax havens (Kuchciak, 2007). On the basis of the literature review of the subject of the thesis, the following conclusions were made. According to the law, the dependent state/territory can be classified as a tax haven by entering it into an international list or upon the entry into force of the relevant act.

Many countries using the concept of tax haven for the purposes of their own law have created different definitions. Liberal regulations, minimal banking supervision, and anonymity are the most important features that allow a given area to be included in the tax haven. These areas are usually characterized by a small area and small population. They are often former or present British colonies. Most tax havens are island territories. The location of tax havens has not changed. Due to the restrictions applied by international organizations, the lists cover a smaller number of territories. An interesting issue is the issue of publishing lists of tax havens.

The large number of published lists prepared by various entitles has little differences. It happened that one of the lists as given state was a tax haven, while on another list, it was omitted. Various criteria for the assessment of a given territory were adopted, and for each entity, something else was important, which led to nonconformities (Rydlak 2017). As a result of the research work, a typology was obtained. As already mentioned, its main purpose and task were to organize the collected data, but also answer the following questions: do selected types of tax havens show regularities in spatial distribution? Are there dominant types? The analysis made allowed to formulate the following conclusions. Modern developed economies are characterized by the dominant role of the services sector in GDP and employment. The share of services in GDP for tax havens amounted to an average of 72%. Areas recognized as tax havens stand out due to their small area—average below 8000 km^2 and a small number of inhabitants not exceeding 1 million people.

Some of the researchers are showing that low fines for disclosed tax evasion in tax havens continue to make tax avoidance attractive to investors (Konrad and Stolper 2016). According to Elsayyad and Konrad (2012), the initiative to reduce the number of tax havens may be harmful, as reducing their number reduces competition between havens that remain active. The remaining tax havens are more reluctant to resign from their function. However, according to Menkhoff and Mieth (2019), efforts to reduce international tax evasion reduce bank deposits in tax havens bring results even by 27.5%.

The problem of selling the citizenship by small microstates is occurring. It is a way to avoid European Union and American sanctions and this is a way for wealthy people from the developing countries to have entered or to invest in the well-developed world.

References

Altuntas F, Gok MS (2020) The effect of COVID-19 pandemic on domestic tourism: a Dematel method analysis on quarantine decisions. Int J Hosp Manage 92. https://doi.org/10.1016/j.ijhm.2020.102719

Artemiuk J (2013) Raje podatkowe w prawie międzynarodowym, Prace Naukowe Wydziału Prawa, Administracji i Ekonomii Uniwersytetu Wrocławskiego, Wydawnictwo: E-Wydawnictwo. Prawnicza i konomiczna Biblioteka Cyfrowa. Wydział Prawa, Administracji i Ekonomii Uniwersytetu Wrocławskiego, Wrocław

Balakina O, D'Andrea A, Masciandaro D (2017) Ank secrecy in offshore centres and capital flows: does blacklisting matter? Rev Financ Econ 32:30–57

Bertram G (2004) The Mirab model in the 21st century, "Changing Islands—Changing Worlds". Islands of the World, VIII International conference. Taiwan, pp 749–781

Białek M (2019) Typology of tax havens in the world—Typologia rajów podatkowych na świecie—master thesis. Univeristy of Warsaw, Warsaw, Faculty of Geography and Regional Studies

Bronner F, De Hoog R (2012) Economizing strategies during an economic crisis. Ann Tour Res 39(25):1048–1069

Byrska D, Borowska A (2017) Techniczne aspekty funkcjonowania rajów podatkowych. In: Pujer K, Danielak W (eds) Zarządzanie rozwojem organizacji w zmiennym otoczeniu. Wydawnictwo Exante, Wrocław

Dahles H, Susilowati TB (2015) Business resilience in times of growth and crisis. Ann Tour Res 51:34–50

Dharmapala D, Hines JR (2009) Which countries become tax havens? J Public Econ 93

Dziedzic T, Łopaciński K, Saja A, Szegidewicz J (2010) Polska gospodarka turystyczna w okresie światowego kryzysu, Instytut Turystyki. Polska Agencja Rozwoju Turystyki, Warszawa

Elsayyad M, Konrad KA (2012) Fighting multiple tax havens. J Int Econ 86:295–305

Errico L, Musalem A (1999) Offshore banking: an analysis of micro- and macroprudential issues. IMF Working Paper', 99/5

Eugenio-Martin JL, Campos-Soria JA (2014) Economic crisis and tourism expenditure cutback decision. Ann Tour Res 44(1):53–73

Foks J (2001) 'Raj podatkowy' jako zagraniczny czynnik w systemie podatkowym [Tax haven" as a foreign factor in the tax system]. Biuletyn Polskiego Instytutu Spraw Międzynarodowych nr 5

Głuchowski J (1996) Oazy podatkowe. Dom Wydawniczy ABC, Warszawa

Grzywacz J (2010) Pranie pieniędzy. Metody, raje podatkowe, zwalczanie. Oficyna Wydawnicza Szkoła Główna Handlowa w Warszawie, Warszawa

Hampton MP, Christensen J (2002) Offshore Pariahs? Small island economies, tax havens, and the re-configuration of global finance. World Dev 30(9):1657–1673

Hansen NS, Kessler AS (2001) The political geography of tax h(e)avens and tax hells. Am Econ Rev

Harmful tax competition: an emerging global issue (1998). OECD, Paris

Hines JR, Rice EM (1994) Fiscal paradise: foreign tax havens and American business. Q J Econ 109(1):149–182

Hufbauer GC (1992) US taxation of international income: blueprint for reform. Institute for International Economics, Washington DC
International Monetary Fund (2008) Offshore financial centers. A report on the assessment program and proposal for integration with the financial sector assessment program. www.imf.org/external/np/pp/eng/2008/050808.pdf. 08 Mar 2019
Jarczok-Guzy M (2016) Zjawisko konkurencji podatkowej w obszarze opodatkowania dochodów korporacyjnych. In: Studia Ekonomiczne. Zeszyty Naukowe Uniwersytetu Ekonomicznego w Katowicach nr 271, Wyd. Uniwersytetu Ekonomicznego w Katowicach, Katowice
Jędrusik M (2005) Wyspy tropikalne – w poszukiwaniu dobrobytu. Wydawnictwa Uniwersytetu Warszawskiego, Warszawa
Kachniewska M (2012) Wpływ kryzysu gospodarczego na metody zarządzania obiektami hotelowymi. Int J Manage Econ 35:61–75
Konrad KA, Stolper TBM (2016) Coordination and the fight against tax havens. J Int Econ 103:96–107
Kowanda C (2017) Afera paradise papers. Czy powinniśmy się tym oburzać? Jak w raju, Polityka Cyfrowa, 6 listopada 2017. www.polityka.pl/tygodnikpolityka/swiat/1726444,1,afera-paradise-papers-czy-powinnismy-sie-tym-oburzac.read. 13 Nov 2017
Kuchciak I (2007) Istota i metody przeciwdziałania wykorzystaniu rajów podatkowych w polskich warunkach. In: Mikołajczyk B (ed) Przedsiębiorstwo a rozwój gospodarki. Acta Universitatis Lodziensis. Folia Oeconomica, Łódź
Kuchciak I (2012) Raje podatkowe w zmniejszeniu obciążeń podatkowych. Wydawnictwo Uniwersytetu Łódzkiego, Łódź
Kurcil-Białecka R, Raj na ziemi (2011) Manager's life. https://kancelaria-skarbiec.pl/pdf/managers-life.pdf
Lipowski T (2002) Raje podatkowe. Charakterystyka i sposoby wykorzystania, Wyd. Ośrodek Doradztwa i Doskonalenia Kadr, Gdańsk
Lipowski T (2004) Raje podatkowe a unikanie opodatkowania. Wydawnictwo C. H. Beck, Warszawa
Mara ER (2015) Determinants of tax havens. Procedia Econ Finance 1638–1646
Masciandaro D (1999) Money laundering: the economics of regulation. Eur J Law Econ 7(3):225–240
Mazur Ł (ed) (2012) Optymalizacja podatkowa. Wolters Kluwer Polska, Warszawa
Menkhoff L, Mieth J (2019) Tax evasion in new disguise? Examining tax havens' international bank deposits. J Public Econ 53–78
Nawrot RA (2011) Szkodliwa konkurencja podatkowa, Wyd. Difin, Warszawa
Okumus F, Karamustafa K (2005) Impact of an economic crisis. Evidence from Turkey. Ann Tour Res 32(4):942–961
Palan R, Murphy R, Chavagneux C (2010) Tax havens. How globalization really works. Cornell University Press
Preuss L (2012) Responsibility in paradise? The adoption of CRS tools by companies domicilied in Tax Havens. J Bus Ethics 110:1–14
Radlak K (2017) Zasadność uznawania Republiki Cypryjskiej za raj podatkowy - wybrane zagadnienia. Roczniki Administracji i Prawa nr XVII.
Ronkainen JK (2011) Mononationals, hyphenationals, and shadow nationals. Citizenship Stud 15(2):247–263
Rosati D (2009) Geneza i mechanizm kryzysu finansowego w Stanach Zjednoczonych. In: Bożyk P (ed) Światowy kryzys finansowy. Przyczyny i skutki. Wyższa Szkoła Ekonomiczna w Warszawie, Warszawa
Runge J (2006) Metody badań w geografii społeczno-ekonomicznej- elementy metodologii, wybrane narzędzia badawcze. Wydawnictwo Uniwersytetu Śląskiego, Katowice
Schwarz P (2011) Money launderers and tax havens: two sides of the same coin? Int Re Law Econ 31:37–47
Sigala M (2020) Tourism and COVID-19: impacts and implications for advancing and resetting industry and research. J Bus Res 117:312–321

Sikorski M (2017) Przeciwdziałanie wykorzystaniu rajów podatkowych. In: Zeszyty Naukowe Państwowej Wyższej Szkoły Zawodowej w Płocku. Nauki Ekonomiczne, Tom 26

Škare M, Soriano DR, Porada-Rochoń M (2020) Impact of COVID-19 on the travel and tourism industry. Technol Forecast Soc Change. Preprint. https://doi.org/10.1016/j.techfore.2020.120469

Starchild A (1993) The tax haven report. How to internationalize your capital for protection and profit. Scope International, Waterlooville

Stylidis D, Terzidou M (2014) Tourism and the economic crisis in Kavala, Greece. Ann Tour Res 44:210–226

Swianiewicz P (1989) Społeczno-ekonomiczna typologia miast i gmin w Polsce, Wydział Geografii i Studiów Regionalnych Uniwersytetu Warszawskiego, Instytut Gospodarki Przestrzennej, Warszawa.

The OECD's project on harmful tax practices: The 2004 progress report (2004). OECD, Paris

Towards global tax cooperation (2000) OECD. Paris

UNWTO World Tourism Barometer 8(1) (2010)

Van Fossen A (2017) Passport sales: how island microstates use strategic management to organise the new economic citizenship industry. Island Stud J 13(1):285–300

Węgrzyn G (2009) Rola sektora usług we współczesnej ekspansji gospodarczej. In: Kryk B, Piech K (eds) Innowacyjność w skali mikro i makro. Instytut Wiedzy i Innowacji, Warszawa

Wójcik Ł (2014) Ile kosztuje obywatelstwo? Paszporty na sprzedaż, Polityka, kwiecień 2014 (www.polityka.pl/tygodnikpolityka/swiat/1577405,1,ile-kosztuje-obywatelstwo.read).

Wotava M (2000) Podatkowe raje i usługi offshore. MOW - Małopolska Oficyna Wydawnicza Korona, Kraków

Zdon-Korzeniowska M, Rachwał T (2011) Turystyka w warunkach światowego kryzysu gospodarczego. Prace Komisji Geografii i Przemysłu 18:116–128

Chapter 4
Selected Political Issues for Public Space—Iconoclasm in Respect of Warsaw Monuments from the 1945 to 1989 Period and Post-soviet Monuments in Ukrainian Cities

Krzysztof Górny and Ada Górna

> *'Burząc pomniki, oszczędzajcie cokoły. Zawsze mogą się przydać'* ('As you demolish your monuments, make sure you spare the plinths – they can always come in handy').
> Stanisław Jerzy Lec.

4.1 Introduction

The erection of a new monument or toppling of an existing one is almost always a political act, though naturally this is especially the case where a monument is of inherently political content. The commemorative function of the monument situated in urban space has been known to Europe since ancient times, and both the unveiling of a new one and its likely removal sooner or later represent activities linking up closely with politics. Monuments have always denoted commemoration and remembrance of past history and they have achieved this through both the material aspect and the symbolism (Forest et al. 2002). And even today, though monuments play quite diverse social roles (Nelson et al. 2003), commemoration, remembrance as well as creating places of social memory remain their key purpose and task (Crang and Travlou 2001). As is rightly noted by Dwyer and Alderman (2008): 'monuments are one kind of memorial text, taking their place alongside a wide range of media designed to facilitate remembering and forgetting of the past'. Thus, monuments are an important research topic of both political geography and cultural geography

K. Górny (✉) · A. Górna
Faculty of Geography and Regional Studies, University of Warsaw, ul. Krakowskie Przedmieście 30, Warsaw, Poland
e-mail: krzysztofgorny@uw.edu.pl

K. Podhorodecka and T. Wites (eds.), *Global Challenges*, Springer Geography,
https://doi.org/10.1007/978-3-031-60238-2_4

as they link current politics and the past as well as place and memory. There is extensive literature on these connections not only in relation to monuments, but also other elements of urban landscape such as street names or architecture (Lowenthal 1961; Charlesworth 1994; Azaryahu 1996; Edensor 1997; Hayden 1997; Leach 1999; Osborne 2001; Johnson 2003; Mitchell 2003; Alderman and Hoelscher 2004; Hoskins 2004; Foote and Azaryahu 2007; Cresswell and Hoskins 2008; Lewicka 2008).

An equally valid remark comes from Koerner (2017), who noted that: 'iconoclasm is perhaps the most extreme response to a thing: instead of ignoring or adoring the thing, instead of moving or modifying it, iconoclasts destroy it'. Iconoclasm was thus a term first pressed into service to describe the process by which and instances in which religious images were destroyed (Gamboni 2013). Later, however, the term came to be associated with the destruction of any work of art—with monuments therefore included among them (Skillingstad 2016).

In the context of the present iconoclasm is deemed to entail the toppling and/or removal of a monument—a process that is actually still pretty common in today's world. It occurs because monuments express power, and thus, they can become the focal point to conflict (Jones et al. 2014). For example, American society would seem to have had a deal of experience in this field recently, as an aggressive debate regularly giving rise to iconoclasm has surrounded monuments in southern states linking in one way or another with the Confederacy (Forest et al. 2019). Equally well-publicised was iconoclasm vis-à-vis the Statue of Cecil Rhodes at the University of Cape Town taking place in 2015 (Kros 2015). A student action entitled *Rhodes Must Fall* then spread to other parts of the world and represented a tiny part of the far-wider dilemma for many societies of 'the Global South', which have had—since more or less the 1950s—to decide whether to protect or annihilate their colonial-era heritage.

The aforementioned iconoclasms were associated with profound change of a nature that was political (decolonisation in Africa) and/or social (the struggle for greater racial equality in the USA and South Africa). As is noted by Kattago (2015), the toppling of monuments is often one of the first signs of regime change; and things were no different in the Central and Eastern European Countries (CEECs) during and after the new 1989–1991 'Spring of the Nations' (Jones 2007). It was at that time that far-reaching systemic changes were made, out of the old Communist system and into democracy, liberty, and free-market capitalism (Fig. 4.1).

The fall of the Soviet Union, as the state exerting dominance over the entire region, would combine with emerging states' steady abandonment of Communist-style politics, governance, and economics to necessitate change in cities' symbolic landscapes, given that these had—over more than four decades of activism—become abundantly outfitted in monuments of overt Communist content. As Jones et al. (2014) notice 'in few places has the politics of landscape been as highly charged as in the former Communist states of Central and Eastern Europe'. Objects of this kind had in fact gone up in cities throughout the Bloc and were intended to exert—and certainly received as exerting—their own kind of symbolic pressure, indeed a kind of violence, against citizens. Unsurprisingly then, the fall of that system led in many cases to decisions to decommunise public space, *inter alia* through the removal of

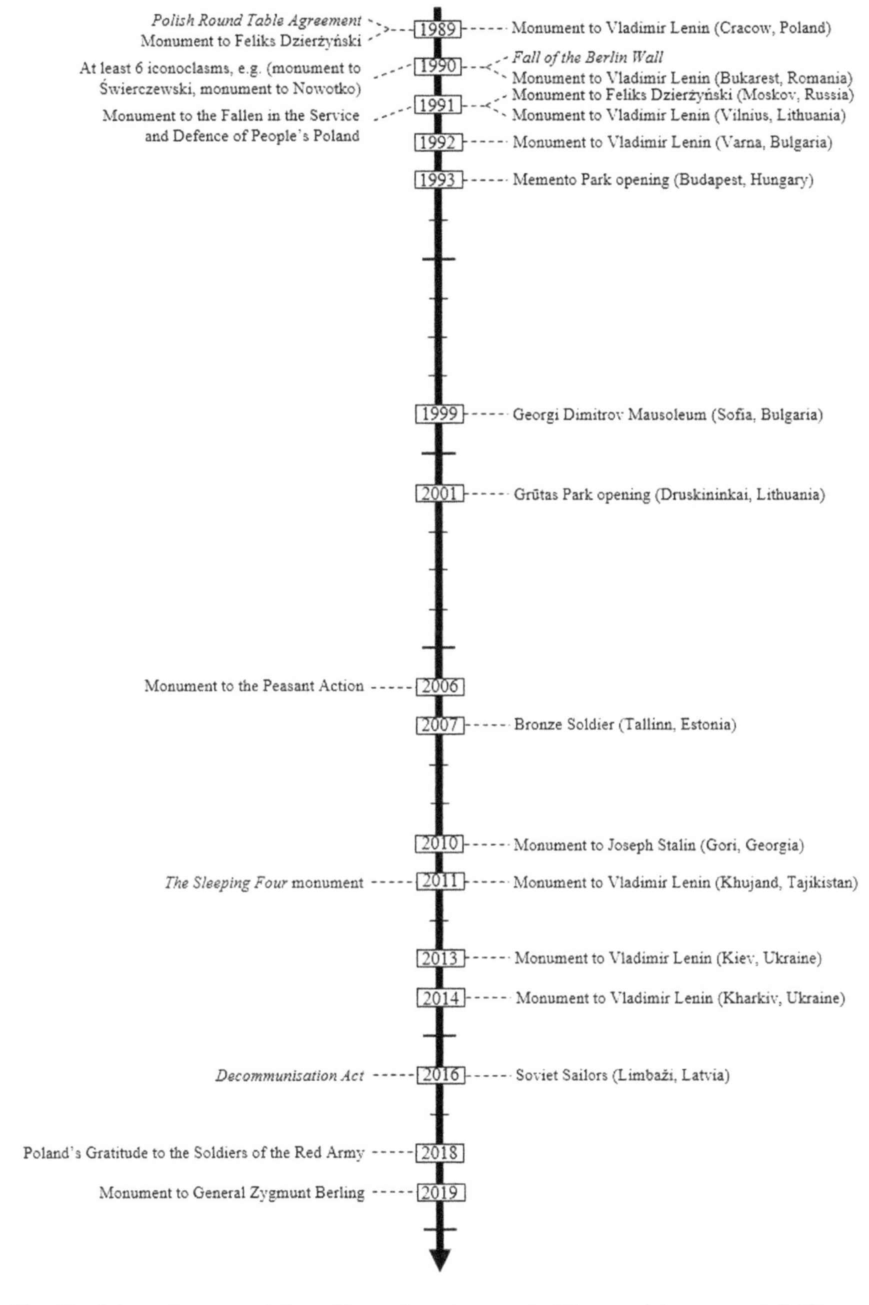

Fig. 4.1 Schematic representation of iconoclasm in respect of Communist monuments in Warsaw (left) and other cities of the old Soviet Bloc (right) in the course of the last 30 years. *Source* Authors' own elaboration

certain identified monuments. This was what happened, for example, in Budapest where the city authorities took away close to 20 monuments of Communist content, including a statue of Engels and Marks (Johnson 1995), Sofia (Ivanov 2009), Tallinn (Forest et al. 2011), and Lithuania's cities (Williams 2008).

However, it is worth recalling that while some cities witnessed the near-surgical removal of all Communist-era monuments, with these at times removed to 'statue parks', as in Budapest (Light 2004), or Park Grūtas near Druskininkai in Lithuania (Kałużna 2018), others practised a 'necessary restoration work' approach that often saw monuments put to use as tourist attractions. This has been the kind of treatment afforded to, for example, Moscow's Lenin Mausoleum (Forest et al. 2002), the Soviet War Memorial in Vienna, or the Monument to Stalin in the Georgian town of Gori which only eventually disappeared in 2010 (Diener et al. 2013). These cases would exemplify behaviour on the part of authorities seeking to preserve historically inflected urban landscapes as a means of accumulating capital (Hoelscher and Alderman 2004). An interesting example related to one of Sofia's Communist-era monuments, which was 'defaced' in 2011, in the sense that particular Red Army soldiers depicted was made over into characters from *Marvel* comics (Stańczyk 2013). This shows that revolution as a process destroys but never fully and not always peremptorily (Czepczyński 2008).

The aim of this chapter is to introduce the reader to the subject of changes that have taken place over the last 33 years in the symbolic space of the Polish capital in order to better understand the ever-changing Warsaw. In the context of recent events beyond the eastern border of Poland, it is of key importance to extend the content of this text to include examples of iconoclasms in Ukraine. The juxtaposition of the changes taking place in Warsaw with those which, in the face of difficult Ukrainian-Russian relations, are occurring in Ukrainian cities is crucial for understanding the processes influencing the transformation of symbolic urban landscapes of Eastern Europe.

This chapter is divided into two parts. The first, much more extensive part constituting the core of this work, seeks to present the very close links between the nature, degree of abruptness, and directions of political shifts taking place in Poland as portrayed by the changes in the symbolic landscape of Polish cities, based on the examples of the monuments erected in Warsaw through the whole 1945–1989 period. The second part discusses the issue of Communist monuments in the post-Soviet reality of Ukraine, a pressing topic in the context of the Russian aggression against this former Soviet republic.

The conceptualisation the authors offer depicts the situation as regards monuments in Warsaw and in Ukrainian cities, as it looks from the perspective of 2022. This is actually a rather delicate moment in this context. As a result of the Russian attack, Ukraine once again faced a choice between East and West. Its authorities decided to bring the country as close to the EU and NATO as possible. This will come at the expense of relations with Russia, which will certainly not return to those from before 2014 in the near future. In turn, in Poland as, with the 1 April 2016 enactment of what has become known generally as 'the Decommunisation Act', the right-wing *Prawo i Sprawiedliwość—PiS* ('Law and Justice') government remaining in power has obliged local governments (including the Centrist administration in place in Warsaw), to free the public spaces in Polish cities of all symbols deemed to propagate totalitarian systems, including above all Communism. The changes in Warsaw are supposed to involve—among other things—the naming applied in cities to streets, parks, squares, bridges, and schools, as well as the physical monuments constructed, in the years 1945–1989, with a view to people, organisations, events, and dates of symbolic relevance to the Communist system being remembered. After six years of 'the Decommunisation Act' and its provisions being in the Polish capital has witnessed isolated incidents of iconoclasm relating to Communist monuments, even as several continue to play their part in shaping the Warsaw landscape. In the face of the war in Ukraine, these actions are increasingly gaining approval of the Polish society, which provided unprecedented support to the Ukrainian cause, especially to the millions of refugees who crossed the Polish-Ukrainian border in 2022. Changes in the symbolic landscape in Polish cities may serve as a tool for demonstrating social unrest and opposition to the actions of the aggressor—Russia.

It is crucial to highlight how urban space can be sensitive to systemic changes, especially those which are thoroughly rebuilding the socio-political and economic system (here the fall of Communism and the Russian aggression). Thus, the chapter also aims to show how urban space, understood as a process of constant change and a system of interconnections, can express public sentiments as well as the mentality of its creators and users. Emphasis is here placed on the iconoclasm as a means of expressing these changes and on the diversity and complexity of their manifestations. It should be pointed out that the monument's life does not end with its removal from space. As the following examples show, it is not only the removal of the monument that counts, but also what happens to it afterwards and what happens to the place where it stood—whether and how the space is redefined. This text is in the nature of a review and presents Polish (Warsaw-based) and Ukrainian iconoclasm in respect of the monuments of Communism taking place from the end of 1989 in Poland and 1991 in Ukraine onwards. The fall of Communism is thus treated as a stimulus to the removal or destruction of monuments: for urban space reflects historical processes and political change, having a landscape that registers events of importance to the society that simultaneously resides in and shapes that space. Poland here serves as an exceptionally interesting and also very complicated case and yet the issue of the decommunisation of public space in the 'country along the Vistula' has thus far been taken up only rarely in international literature—in some contrast to the corresponding situation in Ukrainian cities and in many other cities of ex-Communist-Bloc states such as Budapest (Foote et al. 2000), Bucharest, Berlin, or even Moscow. This may reflect a fact pointed to by Solarz (2014), that Polish political geography was basically non-functional in the whole 1939—1989 period, with only a very slow renaissance post-1989. This is probably why analyses of the relationship between politics and space in Polish cities have only been pursued rarely.

The research was conducted using mainly qualitative methods. In the case of Poland, an in-depth analysis was carried out through, first of all, an inventory of politically marked Warsaw monuments erected during the periods of People's Poland (1945–1989) and post-Communism (1989–present) which was made. Then, a detailed analysis of scientific literature, newspaper articles, and media reports regarding acts of iconoclasm in the period after 1989 in Warsaw was carried out. An inventory was also made of those Communist monuments that survived decommunisation process and are still present in the urban space of Warsaw. Using exploratory and classification approach, we attempt to group the inventoried iconoclasms by their character and show the relationships between the changes in the symbolic landscape of Warsaw and political changes that occurred in post-Communist history of Poland. Simultaneously, we refer to similar examples from other cities in the region. In the Ukrainian case, the scientific literature and media reports were studied to briefly outline the post-Communist and anti-Russian iconoclasm in the country.

4.2 Warsaw's 1945–1989 Monuments

The story of Communist-era monuments in Poland's cities dates back to the end of the World War II, when the Eastern Front moved steadily across the territory of today's Republic. Through its activity and considerable effort, the Red Army was able to push the forces of the Third Reich back westwards, up to the moment Berlin was taken in May 1945. The Soviets' successful occupation of Polish lands was indeed achieved at the cost of the lives of hundreds of thousands of Red Army soldiers. Indeed, close on 600,000 military personnel of the USSR lie buried at cemeteries in different parts of Poland.

It was at such sites that the first monuments to the Soviet Army went up. As a result of what was established by the 'Big Three' of Stalin, Churchill, and Roosevelt, areas of Europe taken by the Red Army in wartime were to pass over permanently into the Soviet sphere of influence. And to retain control over the land conquered (including the entire area of present-day Poland), Stalin decided on the immediate establishment of puppet Communist governments whose tasks would be to safeguard the interests of the Soviet Union throughout Central and Eastern Europe. This was also the stage at which Soviet propaganda activity was commenced with—one of the aims being to glorify the actions of the USSR in taking on the might of Hitler's Germany. Just one focus of this propaganda entailed the raising of monuments to recall the martyrdom and devotion of Red Army soldiers as they liberated the Polish lands. A further aspect related to the brotherhood in arms of the USSR's soldiers and that part of the Polish Armed Forces acting in their support. What was therefore being put in place at that time was an image of the Red Army as an ally and liberator (Ochman 2013). An approach whereby monuments symbolise strength and coercion at one and the same time was and remains typical for totalitarian systems (Donohoe 2002).

The first Warsaw monument to be erected after the Second World War was one unveiled in Praga District (right-bank Warsaw) in 1945—to commemorate Polish-Soviet Brotherhood-in-Arms and hence recall the joint effort to fight Nazi Germany. Praga had in fact been liberated by the Soviet Army as early as in 1944, and so it needs to be recalled how Stalin appears to have deliberately halted his offensive at the River

Vistula, awaiting the bleeding-out of the (left-bank) Warsaw Uprising launched by Polish partisans against the Germans at that point. Against that background reality, a second monument—dating from 1946—related to Poland's Gratitude to the Soldiers of the Red Army and was located in Skaryszewski Park. This is also in the eastern part of the city near the place of burial of several dozen fallen soldiers of the Red Army.

The start of a further period of putting Communist monuments in place in Warsaw can be dated to 1951, when what is today Bankowy Square became the location for a monument to Felix ('Iron Felix') Dzerzhinsky (1877–1926)—a man of Polish origin regarded as one of the most brutal and merciless politicians active at the time the Soviet Union came into being. From that time onwards, the city was almost exclusively involved in remembering by means of monuments such persons associated with Poland's Communist movement as Marceli Nowotko, Julian Marchlewski, Marcin Kasprzak, and Karol Świerczewski.

A further phase of enhanced monument-building in Warsaw coincided with the period following the introduction of Martial Law across Poland (which remained in place in the years 1981–1983). The increased number of Communist-themed monuments at that time came in the aftermath of society's resistance to the Communist government led by General Wojciech Jaruzelski. The monuments of this period recalled the life of Zygmunt Berling, a General who had died in 1980, as well as the formation commanded by him known as the First 'Tadeusz Kościuszko' Infantry Division, which operated within the Red Army during the War. Work on both of these monuments came to fruition in 1985. An act of particular symbolic significance prompted by Varsovians' involvement in protests against the then authorities was the July 1985 unveiling of a Monument to the Fallen in the Service and Defence of People's Poland—the largest and heaviest such work ever to take its place in Poland's capital city. 1986 then brought the unveiling of a Monument to Władysław Gomułka leader of Poland in the years 1956–1970 (who had died in 1982).

Worth adding at this point is the fact that the process of erecting monuments in Warsaw in the years 1945–1989 was not linked solely with the recollection of Communist-related persons and events. For the city's double tragedy in the form of the failed 1944 Warsaw Uprising and the associated (subsequent) razing of what was left more or less district by district on Hitler's direct orders resulted in a situation immediately post-War whereby a Warsaw now 85% destroyed (Andrusz 2008) retained hardly any monuments at all, and most especially where these had in any way been associated with Poles' national identity.

Such a national motif actually played a particularly major role in the days of core 'Socialist Realism' (meaning 1949–1955/6 in Poland's case). Thus, alongside works of the most-typical kind, notably the centrally located and entirely dominating 1955 Palace of Culture and Science with its numerous associated pieces of sculpture (Zaborowska 2001), there was an effort to rebuild monuments to certain figures from the past most worthy and deserving from the point of view of Polish culture. Thus, the Copernicus Monument returned to Krakowskie Przedmieście Street as early as in 1945, while even King Zygmunt III was put back on his Column in Castle Square in 1949. Then in 1950, President Bolesław Bierut himself unveiled the monument to Polish national 'Bard' Adam Mickiewicz, while 1958 brought the completed renewal and unveiling of the former Secessionist-style Monument to Fryderyk Chopin in Łazienki Park (the original of which dated back to 1926). In subsequent years, further 'politically neutral' monuments came on the scene—commemorating female writers Eliza Orzeszkowa (1958) and Maria Konopnicka (1966), male writers Bolesław Prus (1977) and Stefan Żeromski (1987), and even composer and inter-War Prime Minister of Poland Ignacy Jan Paderewski (1978).

4.3 The 1989 Systemic Changes and the Evolution of the Polish Political Scene

On 4 June 1989 was Poland's first and actually partially free parliamentary elections since the days of the Second Republic. These followed on from the so-called 'Round Table' Talks. Though formally the change of name from People's Republic of Poland (*Polska Rzeczypospolita Ludowa, PRL*) to (Third) Republic of Poland only took place with the passing of amendments to the Constitution dated 28 December that year, the June elections to the 'Contract Sejm' (lower House of Parliament) are what are taken to symbolise the fall of Communism in Poland—and thereby also the beginning of the end for that system across the whole of Central and Eastern Europe. The Polish achievement was a peaceful change of regime standing in some contrast to the rather bloody Romanian Revolution (of December 1989), and also far less of a media event then the hugely symbolic and internationally renowned fall of the Berlin Wall of November 1989.

The fall of Communism in Poland was associated with the rise to power of people associated with the 'Solidarity' (*Solidarność*) camp. The first non-Communist Prime Minister was Tadeusz Mazowiecki, who took up his post on 24 August 1989. The following year featured further democratic elections that saw the first new-style Presidency in the free Poland taken up by Lech Wałęsa—as leader of the workers' movement, Solidarity legend and 1983 winner of the Nobel Peace Prize.

However, precisely because the transfer of power to the anti-Communist opposition had taken place in this peaceful way, and by consent, the ongoing political change did not entail full or abrupt decommunisation, and nor did it extend to the broad-scale screening and vetting of individuals' past involvements (Killingsworth 2010). For example, there was even neglect when it came to the enactment of regulations (of the kind that did appear in the Czech Republic) limiting access to public positions in the case of Communists.

This chapter is not a place for any more-profound analysis of the causes and effects of moves of the above kind, but it is worth mentioning that the series of drastic economic reforms proposed by the new authorities in free Poland were by no means fully successful (Bernhard 1996; Kołodko 2009), while splits in the once-monolithic *Solidarność* camp sufficed to ensure the return of 'post-Communists' to power in Poland as early as in the mid-1990s. In 1995, Lech Wałęsa was defeated in the race for the Presidency by Aleksander Kwaśniewski, a former Minister in the Communist Governments of 1985–1989. Even before that—as soon after the systemic transition as in 1993—elections to the Second-Term Parliament of the 3rd Republic had handed power to the 'Democratic Left Alliance' (*Sojusz Lewicy Demokratycznej – SLD*), a grouping bringing together people formerly associated with 'The Party' (i.e. the Polish United Workers' Party or *PZPR*) (Jasiewicz 2008).

Through the following decade to 2005, successive governments alternated between *SLD* and groupings arising out of *Solidarity* (Zubek 2008). At that time in Poland—as Solarz (2012) notes—the identity of a Polish government as either 'democratic or socialist' was found to be a matter of no significance to 42% of citizens polled.

However, the presidential and parliamentary elections of 2005 made it clear that Poland's political scene was beginning to polarise in a new way. A now (finally) marginalised post-Communist option ceased to play a leading role as it made way for the two new forces known as 'Law and Justice' (*PiS*) and 'Civic Platform' (*Platforma Obywatelska* or *PO*). Though both parties had (and have) a clearly Solidarity-related pedigree, they have come to differ very markedly in their approaches to the fallout from the post-1989 period of transformation. Part of that has also obviously related to the way in which accounts with Poland's past have (or have not) been settled. Of a centrist political complexion, *PO* became some kind of further propagator of the already-introduced 'evolution rather than revolution' political model, while *PiS* has sought to undermine and discredit the achievements of the first democratically elected governments, regarding 'decommunisation' (as broadly conceived) as a matter never finished with. This viewpoint culminated in a 2004 suggestion that a Fourth Republic of Poland can be established in place of the Third one (Ochman 2010).

The two aforementioned parties have come to dominate Poland's political scene over the most-recent 17-year period and have alternated in terms of their time in power. It is not therefore hard to deduce that the returns to power of *PiS* have been associated with intensified decommunisation activity in public space, with a kind of peak reached in 2016—with the adoption of a new Act regulating ways in which persons, events, and dates associated with *PRL* might be recalled.

4.4 Iconoclasm and the 2016 'Decommunisation Act'

One of the motifs behind the authors' addressing of issues contained in this chapter was in fact the Polish Parliament's 2016 adoption of the said Act relating closely to decommunisation, notwithstanding a full title that may be translated as:

> The Act of 1 April 2016 prohibiting the promotion (or propagation) of Communism or any other totalitarian political system by way of the names of organisational units, auxiliary units at local-authority level, buildings, or other objects and public-utility installations, as well as monuments.

The adoption of this piece of legislation came in the aftermath of the aforementioned change on Poland's political scene following the 2015 victories by *PiS* in both the presidential and parliamentary elections. A *Law & Justice* party of right-wing profile (in this dimension at least) commenced its 2015 campaigning with slogans propounding the final conclusion of a process to decommunise Poland that had admittedly begin in 1989, but was viewed by that party's politicians as unsuccessful, imprecise, and selective at best. They asserted that a final completion of the process would make it clear who had in reality fought (and was still fighting) against Communism and who had reached some kind of accommodation with the Communists. This was also then a criticism of the fundamental apparent achievements ascribed to the Round Table talks and Poland's peaceful systemic transformation.

Among other things, the Act as made ready was to put a final end to the question of elements present in the landscape of Poland's towns and cities that still recalled (or paid homage to) the Communist era, with a list of objects slated for removal even designated (Stryjek 2018). This related above all to still-existing urbanonyms, as well as a number of physical monuments. It was considered that 500 of the latter of Communist content had come into existence in Poland (Czarnecka 2015), with around 300 remaining at the time the Act came into force.

The consequence of the enforcement of these provisions was thus the removal of many monuments in Poland, including the one honouring Gen. Karol Świerczewski who had remained in place in Jabłonki in southern Poland (as the site of the General's murder in 1947), the Monument of Gratitude to the Red Army in Legnica (where that Army had continued to be based through to 1993), and the monument of similar designation located in the Upper Silesian city of Katowice. However, a specific feature here was the way that monuments founded as part of national policy—and now to be removed as part of national policy—were nevertheless to be the responsibility of local authorities. Opposition on the part of the latter was then mainly an economic matter, as removing a single monument might well cost something in the order of 10,000+ euros up to several tens of thousands of euros. Some local authorities—and particularly those in the hands of parties opposing *PiS*, have thus seemed to drag out the process of by which public space is being decommunised.

4.5 Iconoclasm in Post-Communist Warsaw

Subsequent sections here will present post-1989 examples of Warsaw iconoclasm in chronological order taking into account the political context associated with political change in post-Communist Poland. As Czepczyński rightly points out (2008), 'landscape revolution goes behind the political transformation'. The events described point to the varied nature of activity engaged in to remove Communist monuments during the times of the Third Republic of Poland.

4.5.1 Iconoclasm Incidentally of a Revolutionary Nature

The momentous year 1989 obviously marked the onset of post-Communist iconoclasm, meaning the first demolition of monuments that had been put up in Warsaw through the 1945–1989 period. First to make its exit from the public space of Poland's capital city was the Monument to Felix Dzerzhinsky, whose demolition was engaged in by the Urban Roads Directorate at provincial level—between 16 and 17 November. To this day, the gesture of iconoclasm this came to represent has been seen as symbolising Poland's liberation from Soviet occupation (Czepczyński 2010). Indeed, it is by far the most recognisable of all the gestures of this type to have taken place in Poland from the time of the 1989 transition onwards. As work with a crane began to remove a pedestal made of concrete and only in fact coated with a thin layer of bronze (not bronze all the way through, as had been believed), that disintegrated in mid-air, with the severed head of 'the Red Torturer' crashing down to the ground from a height of several metres—and with all taking place before the very eyes of a crowd of cheering Varsovians. As Grzesiuk-Olszewska (2003) wrote: 'the large and highly enthusiastic crowd of people that had assembled went on to sing the national anthem, with the whole event shown the same evening on the main evening news bulletin'. Western media, including the BBC, also got hold of the footage, and presented it.

The sculpture of 'Iron Felix'—a person much hated by Poles—was by Zbigniew Dunajewski and had been unveiled on 22 July 1952 in the Square named after Dzerzhinsky himself, in recognition of the 25th anniversary of the death of this shaper of Soviet Union statehood, and first head of the *Cheka* (Gladsky 1986). The Monument was inaugurated on the key national holiday back in the days of People's Poland by President of Poland Bolesław Bierut, accompanied by Foreign Minister of the USSR Vyacheslav Molotov and Marshal of the Soviet Union Ivan Konev. This was then the first monument to an activist in the workers' movement erected in Warsaw.

Nevertheless, the symbolic and much talked-about nature of events in central Warsaw in November 1989 was in fact a chance happening and spontaneous action. The toppling—later to become known as 'the fall of Dzerzhinsky', or indeed the symbolic 'fall of Communism in Poland', went unannounced and unheralded, and it drew greater crowds merely because it lasted two days. The city authorities had sought to avoid any more major publicity, hence official motivation for the removal of the monument being given as the need to build Warsaw's first *Metro* line due to pass right under the Square (Zaremba 2018). In the event, the figure's fall was a reflection of the fact that people were under informed in such spheres and lacked necessary know-how. In the end, it would be the first of many iconoclastic acts, but the only one to manifest itself as revolutionary in any real sense.

Dzerzhinsky was supposed to have 'gone quietly' from Warsaw, but in the event he was overthrown before cheering crowds. And although the systemic change that was then taking place in Poland was not revolutionary, or spontaneous, or ostentatiously media-savvy, the iconoclasm under discussion here was all of those things. And, as has been noted, that was more or less by chance. Nevertheless, the reaction from members of the public gathered in the Square was open and honest. It revealed the mood in society and certain needs linking up with the abrupt end that had been put to the era of Soviet domination over Poles. One film clip from the day the monument was toppled sees a (female) Warsaw inhabitant share her opinion to the camera: 'it is a joy for me that those bastards are going away from us (…). We are returning to Europe! It's the most beautiful day of my life!'. Similar events in fact took place in Moscow in 1991, when the (ultimately abortive) coup launched against Gorbachev saw the opportunity seized to remove the Monument to Dzerzhinsky located in front of the Lubyanka Prison (Levinson 2018; Harvey et al. 2015).

Learning from the example of the ad hoc revolutionary iconoclasm represented by the Dzerzhinsky Monument, the Warsaw authorities ensured that subsequent demolitions took place in a far more discreet and even surreptitious manner. Thus did further Warsaw monuments from the Communist era make their disappearance—without fanfare, or indeed secretly.

Almost 12 years on from 'Dzerzhinsky's Fall', 29 September 2001 witnessed the appearance of a new monument in *Plac Bankowy* (the Square had become known as Bank Square instead of Dzerzhinsky Square from the time of removal of the old monument). A statue with a likeness of key Romantic-era Polish poet Juliusz Słowacki (1809–1849) was positioned more or less exactly where Dzerzhinsky had stood.

4.5.2 Quiet Iconoclasm

The early 1990s brought further instances of the decommunisation of Warsaw's urban space, with the city's authorities taking away monuments honouring Gen. Karol Świerczewski, Marceli Nowotko, Julian Marchlewski, Marcin Kasprzak, and Władysław Gomułka, as well as those who Fell in the Service and Defence of People's Poland.

General Karol Świerczewski (1897–1947) was actually a Warsaw-born Communist activist and ultimately a high-ranking military office in three different armies. Marceli Nowotko (1893–1942) was the first leader of the Polish Workers' Party, and one of the leading activists of a Communist movement reviving in Third Reich-occupied Poland. Julian Marchlewski (1866–1925) was one of inter-War Poland's leading Communist activists and hence a co-founder with Lenin of the *Communist International*. Marcin Kasprzak (1860–1905) was another Polish activist in the workers' and socialist movement in the Wielkopolska region. Władysław Gomułka (1905–1982) was a Polish Communist politician who was the de facto leader of his country from 1956 through to 1970, as Secretary of the Central Committee of the Polish United Workers' Party (*PZPR*).

The above biograms make it clear why these Communist activists were among those honoured by the many monuments that went up in the Warsaw of that period. In 1957, the 10th anniversary of the death of Świerczewski was commemorated in

Warsaw's Śródmieście district by a bust that was the work of Alfred Jesion (Grzesiuk-Olszewska 1995). The same district of the city also gained monuments to Marchlewski and Nowotko in 1968, while in Wola—a little further west—another monument to Świerczewski was installed in front of the Precision Products Factory, with the artist in this case being Gustaw Zemła (Grzesiuk-Olszewska 1995). 1975 brought the 115th anniversary of the birth of the aforementioned Kasprzak, so a monument to him created by Edmund Matuszek was likewise unveiled. A further monument to Nowotko was placed before the Factory named in honour of him in 1972, while one remembering Władysław Gomułka was put up in front of Warsaw's huge car factory on 6 November 1986. Four years earlier, the Bulgarian government had made its gesture of friendship by gifting a work honouring Georgi Dimitrov.

Not unnaturally, the change of political and economic system brought a determination that all of these monuments would be removed from the cityscape of what was now the capital of a free Poland. In essence, there was no place—and certainly no welcome—in Warsaw's public space for figures unambiguously associated with Communism as a movement. However, learning the lesson that the scenes surrounding Dzerzhinsky's Monument had taught them, the city authorities sought to engage in iconcoclasm rather 'on the quiet'. The above figures made their disappearance between 1990 and 1991, but it is actually now hard to determine on what precise date they were removed. Certainly journalists failed to document the actions, and they seem to have been neither photographed nor filmed. This was then a discreet process, of a top-down nature, carefully orchestrated and without any fuss or fanfare. However, an exception not really providing for anonymity involved the Nowotko Monument located outside the plant bearing his name. This action was a grassroots initiative, however, though it took place with a democratic say-so, given the holding of a vote among employees—who came out 65% in favour of removal (Zaremba 2018).

The first wave of enhanced—if surreptitious—decommunisation of Warsaw's symbolic urban landscape (coinciding with the early 1990s) did not end with the erasing of the 'outstanding' activists of Communism, as 1991 brought the removal of the monument to those who 'Fell in the Service and Defence of People's Poland' (which had gone up shortly after the Martial Law period—on 20 July 1985). This was the largest post-War monument in Warsaw, but it also proved to be the one of most-limited duration, and the last ever to seek to promote Communism as a philosophy. It weighed no less than 60 tonnes, was up to 16 m tall, and came in 483 separate parts. It presented workers (naturally), but also members of the *Milicja* (Communist-era police force), soldiers, peasants, and members of the intelligentsia all raising the white eagle that takes pride of place on the national emblem of Poland (Grzesiuk-Olszewska 1995).

This work had been unveiled in the presence of the state's highest authorities, including of course First Secretary of the Party General Wojciech Jaruzelski. This was a second monument—other than that paying homage to Dzerzhinsky—whose place was actually taken by another; for it was in 2010 that this site came to play host to a likeness of Tadeusz Kościuszko, a participant in the Polish-Russian War of 1792 and the American War of Independence 1775–1783. The Monument was therefore unveiled by President of Poland Bronisław Komorowski, as well as the then US Ambassador in Poland William Heidt.

4.5.3 Random Iconoclasm

Following a period of the intensified demolition of monuments of Communist content (1989–1991), Warsaw reached a moment at which—as Zaremba (2018) put it succinctly: 'the time had come to put up new monuments, rather than topple old ones'. If a monument disappeared, that was then a kind of chance happening. This is for example what occurred with the Monument to the Peasant Action, otherwise *Światowid*, which was by Stanisław Sikora and had been unveiled in 1968 in Warsaw's centrally located Śródmieście District (Grzesiuk-Olszewska 1995). The Monument was placed besides the building housing the Polish Peasant Party then known as *Zjednoczone Stronnictwo Ludowe* (*ZSL*). It had been paid for by that party (which had come into existence in 1949 as a satellite grouping of *PZPR*). As it featured four faces looking in the four main directions (in the same way as pagan statues of Sviatovid dating from the eleventh century), it earned the above nickname. Interestingly, the faces were two real-life ones—of Polish peasants Bartosz Głowacki and Michał Drzymała; as well as two of an abstract natures, representing a *Green Cross* nurse and a partisan of the Peasant Battalions. These references were thus to organisations arising during World War II to continue the fight against the occupying Germans.

As this last monument aroused no particular controversy (unlike others from the Communist period), there was no plan for it to be removed during the first wave of decommunisation. However, the authorities of *Polskie Stronnictwo Ludowe* (*PSL*—as the successor to *ZSL*) made a 2006 decision to sell the plot that included the party's seat to a private developer. Nobody apparently remembered about the monument until the firm *Dom Development* had begun to dismantle it. Interestingly, those standing up in defence of the work at that late stage included, besides *PSL* politicians, Warsaw inhabitants and art critics. This last monument was thus transferred to the Museum of the Polish Peasant Movement, located some 300 km away from Warsaw (Zaremba 2018). This was then a random iconoclasm, offering one of the few examples that had no distinct political basis.

4.5.4 Mature or Secure Iconoclasm?

Another important Warsaw monument of Communist subject matter was removed in 2011. This was the Monument to Polish-Soviet Brotherhood-in-Arms, and it stood in *Plac Wileński* in right-bank Warsaw's North Praga District. It had been unveiled in the presence of President Bolesław Bierut and Prime Minister Edward Osóbka-Morawski on 18 November 1945, as Warsaw's first post-War monument. While this featured likenesses of army personnel from both Poland and the USSR, it was topped by three Red Army soldiers running westwards and—*inter alia*—throwing grenades, while the four figures of Polish soldiers—located below—appear to be more or less standing guard. The bronze work was cast in Berlin, and the monument features the inscription loosely translated as: 'In praise of the heroes of the Red Army as Brothers in Arms, who gave their lives for the freedom and independence of the Polish Nation, this Monument was raised by the inhabitants of Warsaw in 1945'.

This Monument resembled that to Dzerzhinsky in meeting its end in connection with work to develop the *Metro* in Warsaw. Construction of the *Dworzec Wileński* Station in the central part of Line II required that the Monument can be removed at least temporarily, though its return to another part of the same Square was foreseen at least initially. In 2011, following the daubing of the work with red paint (by those advocating its permanent removal), as well as the later dismantling of the figures, the Monument was sent for conservation measures to be carried out, while a new plinth was being prepared. However, while a group of local residents made their opposition to further work plain (Śmigiel 2012), a poll of Warsaw inhabitants carried out by *Gazeta Wyborcza* revealed that 55% of respondents claimed to be for the Monument's return (with a majority regarding it as a permanent landmark and point of reference in the place of their birth. Twenty-three % expressed the view that the work should be disposed of, while 22% apparently had no opinion on the matter (Szpala 2013). In essence, the inhabitants of Praga District seemed to have managed to reinterpret'the Sleeping Four' (as it was termed typically), depriving it of its historical and commemorative role and instead bringing it into their own private domain, as one aspect or another of their personal experience (Jałowiecki 2008).

However, despite this acceptance of the Monument by the local community, the ultimate (February 2015) decision of Warsaw City Council was that its reinstatement should be resigned from (Wojtczuk 2015). This was (deemed to be) a decision reflecting practical problems with the unfavourable nature of the new location proposed (too close to buildings), as well as a lack of agreement regarding any entirely new location. Moreover, a negative opinion as regards any reinstatement had been given by the Council for the Safeguarding of Remembrance of Fighting and Martyrdom). Also avoided through the decision arrived at were potential costs of having the Monument cleaned—perhaps repeatedly—were it to receive further doses of red paint.

It was in this way that the long (more than 4-year) debate on the return of the Monument gave rise to the least-controversial decision, and one that at the same time put an end to the matter (Ochman 2018). The determination that consent for its return was lacking thus resulted in part from the public mood and in part from practical—as well as political—considerations, and it represented a solution that acted to silence conflict. Though most Warsaw inhabitants could see a continued place for the Monument in the city's landscape, their ability to prevail over staunch opponents, or those offering no opinion, was not convincing. For their part, the radicals were ready to protest and go on profaning the Monument; and they made that fact quite plain. Just several months prior to the parliamentary elections that brought *PiS* to power (while depriving *PO* and the *PSL* thereof), the Warsaw authorities (of that same *PO*) sought to distance themselves from accusations that they were propagating the memory of—or somehow maintaining in symbolic form—Soviet domination over Warsaw by way of the Monument to Brotherhood-in-Arms. The decision they took was therefore intended to be safe, in the context of both the public mood and the possibility of political gain.

4.5.5 Legitimised/Legal Iconoclasm

A further example of iconoclasm in respect of a Communist-era monument in Warsaw was furnished in late 2018, when an end was put to the Monument of Gratitude to the Soldiers of the Red Army, which was located in Skaryszewski Park in the eastern part of Warsaw (Gawlik 2018). That Monument had been unveiled in 1946 at the place of burial of 26 soldiers of the Red Army who had fallen in the 1944 Battle for Warsaw. Its removal might have fallen foul of the provisions of a 1994 agreement concluded between Poland and Russia in regard to the graves of victims of the World War II. However, in 1968 the bodies of the soldiers referred to had in fact been exhumed and transferred to the larger place of burial for Soviet soldiers in Żwirki i Wigury Street (left-bank, western Warsaw). In this way, the Monument had ceased to constitute a military cemetery. The decision that the Monument should be removed came from *Zarząd Zieleni* (the Board with responsibility for green space in Warsaw), and permission was also granted by the Conservator of Monuments at provincial (Voivodship) level. This controversial monument, also many times daubed in red paint or even faeces, was thus dismantled and taken away.

The initial assumption had been that the statues involved would be transferred to the aforementioned cemetery for Soviet soldiers, in a way that would have paralleled the treatment of, for example, the Bronze Soldier of Tallin, in 2007 (Smith 2008; Kattago 2009). Ultimately, however, the 2018 dismantling of the sandstone monument ended in its sustaining major damage. The idea to take it to another place in Warsaw was thus abandoned, with only the several surviving plaques transferred out to the Museum of the Cold War located at Podborsko in northern Poland. This was then the first Warsaw monument to be dealt with following the entry into force of the 2016 'Decommunisation Act', denoting that the decision on its removal gained legitimisation from the law in force.

4.5.6 'Social' Iconoclasm

Exemplifying iconoclasm as the result of an organised public action is the case of the Monument to General Zygmunt Berling. Created by Kazimierz Danilewicz, this had been erected in 1985 close to one of Warsaw's main arteries, called *Trasa Łazienkowska*. The Monument resembled the man—who commanded the First Infantry Division (later First Mechanised Division)—in evoking extreme emotions among Warsaw inhabitants. Berling (1896–1980) was indeed seen as a national hero and commander by a part of society, given his August 1944 order allowing soldiers to cross from right-bank to left-bank Warsaw to help Varsovians fighting in the Warsaw Uprising. A landing was carried out, as the inhabitants of left-bank Warsaw were actively taking the fight to the Nazis, but a Soviet Army that had already 'liberated' right-bank Warsaw followed Stalin's orders and stopped the advance, not extending support to the Uprising and patiently awaiting the bloodletting on a grand scale that the continued Polish-German fight was bound to, and did, denote (for the former in particular). While the action Gen. Berling had ordered came to nought and indeed ended in disaster (with considerable loss of life), it came to symbolise in the eyes of many a genuine attempt to assist with the liberation of the Polish capital from under German Occupation.

On the other hand, Berling has also to be seen as a traitor, given his well-documented cooperation with the Communists. Nevertheless, this had in fact begun in October 1939, with Berling's arrest by the NKVD and transfer to a prison camp in Ukraine. It was there, under such circumstances, that he consented to cooperate with the Soviet's power apparatus. Sure enough, as the Anders Army was taking shape, Berling elected to remain in the USSR and establish instead the aforementioned Polish military formations within the Red Army framework. For that, the Polish Command accused him of treason in respect of the Polish State and handed down an *in absentia* death sentence. It was then as part of his cooperation with the Soviet Army that—from 1943 on—he took up the command of the First Polish 'Tadeusz Kościuszko' Infantry Division, *inter alia* seeing action at the Battle of Lenino. The next step—in 1944—was for Berling to lead what had by then become the First Polish Army in the USSR.

The conflicting sentiments as regards the General translated into differing public attitudes to the monument presenting him. But it is also worth mentioning at this point that the very appearance of that monument was not free of its specific symbolism—real or apparent. For the likeness of Berling, carved in white marble, goes down to the knees only! Since the rest of the statue's legs are, as it was, encased in the red-sandstone 'overshoes' represented by the plinth, the General seems to be entirely prevented from moving west, despite this being the direction the figure would clearly like to head in. Urban legend thus has it that the statue's author was seeking to convey an 'immobilisation' of Berling imposed by the commanders of the Red Army, for which he had volunteered.

For all that, the Monument's almost 35-year lifespan saw it gain regular coats of red paint, while the plinth from time to time acquired graffiti to the effect that this figure was 'KGB', 'traitor', or 'agent of Stalin' (Jałowiecki 2009). Its presence, treated as a manifestation of Communist values being propagated, came to be regarded as in contravention of Polish law. However, its removal was never actually achieved, thanks to intervention on the part of its advocates, mostly *Democratic Left Alliance* activists.

Only with the aforementioned 'Decommunisation Act' of 2016, there was a clear and authoritative position as regards the status of the Monument, which was thus (in December 2019) to be transferred to the Store of the Museum of Polish History, which also in fact houses the aforementioned Monuments to Dzerzhinsky and to Polish-Soviet Brotherhood-in-Arms. However, work carried out on 4 August 2019 led to damage, and that was followed by the toppling of the monument from its plinth. The perpetrators have not been identified so far, but far-right circles, *inter alia Confederation for an Independent Poland* activist Adam Słomka, have termed this iconoclasm 'societal removal of a totalitarian monument to a traitor, by opponents of Communism'. The bottom-up 'societal' action thus came in advance of the implementation of a decision from head of Mazowieckie Voivodship (*Wojewoda*) Zdzisław Sipiera (of *PiS*), offering an opportunity for a protest in front of the destroyed monument to be mounted by a group of Polish flag-carriers, among whom one is photographed pretending to stamp on the head of the toppled General Berling.

This iconoclasm pursued in relation to a monument anyway slated for transfer from its location in the same year may thus be seen as more of a manifestation of right-wing, anti-Communist views, than as a desire to see the object removed from Warsaw space. And first and foremost, it may be regarded as an act of vandalism that breaks the law.

The situation with the Berling Monument offers an extreme example of contradictory memory (Jałowiecki 2009) and in some way reflects the mobilisation through radicalisation of public opinion, in regard to the engagement in decommunisation achieved post-1989.

4.6 Non-iconoclasm or no Iconoclasm as yet

More than a decade ago it was noted in reference to post-socialist cities that 're-production of meaning in the urban landscape is definitely not completed' (Czepczyński 2008). To be mentioned here are several monuments in Warsaw that came into being in the 1945–1989 period, carry undoubted political content, and yet—for one reason or another—have (as yet) escaped removal.

The story related by three of the above links up with another already non-extant monument (i.e. that to Gen. Berling). The monuments in question are a 1963 one paying tribute to the soldiers of the First Army, a 1985 monument to the Kościuszko Infantry Division—actually a subdivision of that same Army and a monument 'In Praise of the Sappers'. The common feature here is that all of these army formations from the Second World War were under the command of General Berling, which is also to say that they were Polish military personnel forming part of the Red Army, given that they either failed—or did not seek—to leave the Soviet Union along with the Army of General Władysław Anders, as established by virtue of the so-called Sikorski–Mayski Agreement signed on 30 July.

These monuments have thus differed from the one to Berling himself, in being far less-immediate or obvious targets for any potential decommunisation. The first reason is that whatever the details, the subjects are Polish soldiers who battled Poland's Nazi-German Occupant. What is more, unlike Berling himself—widely regarded as a traitor and active collaborator with the NKVD—these were simple people, ordinary soldiers, who often just 'missed the boat' when it came to the mobilisation of Polish forces masterminded by Anders.

A still-less-clear situation is that relating to the story of the third 'Sappers' Monument, which is by the Vistula not far from the place once known as the Czerniakowski Bridgehead. Remembered here is a landing by Polish contingents under Berling's command and *inter alia* made up of Engineers ('Sappers') whose task was to build a pontoon bridge across the Vistula. The point here is that the propaganda machine associated with the People's Army had been at work to try and 'paper over' the decision to extend no assistance to the 1944 Warsaw Uprising taking place just across the river in left-bank Warsaw. The decision to raise this monument serves to emphasise how much assistance was offered (and how much distinction might be owing) to Polish military formations within the Red Army. It is here worth making clear that it was on exactly the opposite bank of the Vistula from the place that Berling's forces made their 1944 landing that, through to 2019, there stood the monument to Berling—that figure 'unable to move', looking in the direction of the fighting units, and even with binoculars in hand.

Despite the fact that the three monuments referred to here recall Polish soldiers' fighting against the forces of the Third Reich, and notwithstanding the way neither the authorities of today's Polish state nor Warsaw's local government refer to a necessity for all Communist-era monuments to be lost from the Polish capital, the public debate on the subject does feature voices in favour of this, most especially when it comes to those distorting reality and glorifying military formations active within the overall structure of the Red Army.

The second of the monuments under discussion is the white-marble sculpture known as 'Vistula and Volga', which was unveiled in 1985. This presents two seated female forms looking each other in the eye, but having lower parts of their bodies transformed into two flowing rivers. While the first of these is Poland's largest river—the Vistula, the second is the most major river in today's European part of Russia (ex-Soviet Union)—i.e. the Volga. This sculpture symbolising the Polish-Soviet Alliance was located near the entrance to the so-called Polish-Soviet 'House of Friendship' which was in operation in Warsaw's city centre prior to 1989. The artistic installation in question (more of a work of sculpture than a monument) augments a *bas-relief* situated on the building that also presents the two rivers, whose waters here mix together freely. An interesting associated story is that this work was actually renovated, cleaned, and re-unveiled not so long ago, and while it does indeed celebrate and symbolise the fraternal friendship between Communist nations—there is no aspect of the public debate on these kinds of subjects suggesting that this monument would have to be taken away.

A further interesting case study relates to the 1962 unveiling—in the city centre—of a statue in remembrance of the insurgents of 'the People's Army' (*Armia Ludowa*), whose task was to engage with the Nazi-German Occupant, as well as pursuing diversionary activity to the benefit of the (Soviet) Red Army during the World War II. As is evident, the authorities in the People's Republic of Poland had long been hesitant about honouring insurgents and participants in Uprisings (not least the hundreds of thousands of Varsovians who fell in the 1944 Warsaw Uprising—among the bloodiest of the many such events in Polish history).

Activity had thus hitherto been confined to the 1957 conferment of the name *Powstańców Warszawy* (Warsaw Uprisers) to a square previously named after Napoleon Bonaparte. However, the early 1960s brought the erection of monument relating to the Partisans fighting for People's Poland—just opposite the main headquarters of The (Polish United Workers') Party. As is clear, this was a selective act of remembrance, given that it referred solely to those formations within the Uprising that had acted in support of the USSR's military actions. And when this monument was subjected to renovation in 2014, the commemorative inscription referring to People's Poland gave way to one reading *Partyzantom walczącym o wolną Polskę w czasie II wojny światowej* ('to the Partisans fighting for a free Poland during World War II'), which thus ensured that all insurgents were from that time on being commemorated. This symbolic instance of decommunisation of linguistic landscape was thus confined to a name-change, with no actual need for more far-reaching iconoclasm.

4.7 Anti-communist Monuments Put up in Warsaw Post-1989

As Communist monuments were removed from 1989 onwards, so Warsaw's space had room for new and alternative symbolic content. New democratic authorities at both state and local levels have devoted the last 3 decades to erecting monuments in memory of a variety of different figures and events. As the decommunisation of existing monuments was progressing, it was also determined that the change of political complexion at the level of the entire state should be made manifest through the putting in place of monuments that would have been unacceptable, pre-1989, to the Warsaw authorities in the days of a *PRL* subservient to Moscow. The following monuments were erected 'in opposition to the old politically, economically and morally bankrupt system' (Czepczyński 2008).

A large group of monuments relates in terms of subject matter to crimes perpetrated against the Polish nation by the NKVD (People's Commissariat for Internal Affairs of the USSR). Foremost among these was of course the Katyn Massacre perpetrated by officers of the Commissariat against in excess of 20,000 Poles, including over 10,000 officers of the Army and Police. Thus 1998 saw the unveiling of a 'Katyn Stone' in Warsaw's Old Town. Just three years previously, finishing touches had been put to the Monument to the Fallen and Murdered in the East, which pays homage to the thousands of Poles who never returned from their forced exile far into the interior of the USSR from 1939 onwards. The plaque on that Monument reads *OFIAROM AGRESJI SOWIECKIEJ 17.IX.1939* ('to the victims of the Soviet aggression of 17 September 1939'). In addition, in 1997, a monument took a place of honour in the Jewish Cemetery in Wola District, commemorating Officers of the Polish Army of the Jewish faith, and thus bearing the description: *Pamięci Żydów oficerów Wojska Polskiego zamordowanych przez NKWD wiosną 1940 r. w Katyniu, Miednoje i Charkowie* ('To the memory of the Jews who were Officers of the Polish Army murdered by the NKVD in spring 1940 at Katyn, Mednoye and Kharkov'). Prior to that, a 1995 unveiling had involved a Monument to Prisoners of the NKVD Camp at Rembertów, while 2010 and 2017 witnessed the appearance of statues presenting General Emil Fieldorf (pseudonym *Nil*), as well as *Rotmistrz* Witold Pilecki, an activist of the Independent Underground during the World War II who was sentenced to death (and duly executed) by the Communist authorities in 1948. Such remembrance by way of monuments of the victims of Communism has now become a popular motif throughout the CEECs (Bartetzky 2006).

The change in geopolitical orientation has also involved the re-enrichment of Warsaw space through the 1991 return of the statue of American politician Edward Mandell House that had previously been located in Skaryszewski Park through to 1951; as well as a bust of Napoleon Bonaparte restored in 2011 to its former location once indeed known as *Plac Napoleona* (but for some time now named after the Warsaw Uprisers as *Plac Powstańców Warsawy*). In turn, entirely new monuments going up were to Charles de Gaulle (in 2005), the Armed Action engaged in by American *Polonia*, and President Ronald Reagan (in 2011). In such a way, it was possible to put clear emphasis on links between Poland's Third Republic and its NATO allies, notably of course the USA (Fig. 4.2).

4.8 Post-Soviet Monuments in Ukraine

On 24 August 1991, Ukraine declared its independence from the USSR, which was dissolved in December of the same year. From the perspective of the present authorities in the Kremlin, who consider the collapse of the USSR a mistake, this was the greatest demographic loss that Moscow suffered. It was also the loss of an important territory rich in resources and fertile chernozems, providing enormous surpluses of grain for export. From the beginning of the new independent republic, the pro-Russian parties played an important role in the making of the future of Ukrainians, especially in the east and south of the country.

In 1991, Leonid Kravchuk, a Communist and former Chairmen of the Supreme Soviet of the Ukrainian SSR, became the country's first president. In 1994, he was replaced by Leonid Kuchma, who pursued a flexible policy of manoeuvring between Russia and the US/EU. The decade of his presidency was a period of the deepening of the divisions between pro-Western Western Ukraine and pro-Russian Eastern Ukraine. The results of the presidential elections in 2005 between Viktor Yushchenko (president in 2005–2010) and Viktor Yanukovych (later president in 2010–2014) are a clear example of the country's spatial division into the east and west. While Yushchenko pursued a policy of rapprochement with the EU and the USA, an example of which was the application to join NATO in 2008, Yanukovych, born in the Russian-speaking Donetsk Oblast, sought a multidimensional rapprochement with Russia. His decision to refuse to sign the European Union-Ukraine Association Agreement in

Fig. 4.2 Map presenting the Warsaw Communist and anti-Communist monuments. *Source* Authors' own elaboration

2013 triggered a wave of bloodily repressed protests called Euromaidan and ultimately led to the exile of the Yanukovych to Russia. After the early elections in 2014, pro-Western Petro Poroshenko became president. The significant weakening of Russian influence on Ukraine's policy after Euromaidan prompted Vladimir Putin to start a war against his neighbour. Annexation of Crimea (in 2014), War in Donbas (2014–2022) as well as Kerch Strait incident (2018) have permanently divided Russia and Ukraine, whose full-scale conventional war, initiated on 24 February 2022 with Russian attack, is in fact a continuation of the conflict that has lasted for 8 years now.

Such briefly outlined political changes in independent Ukraine have an impact on the symbolic space of its cities, towns, and villages. Seven decades of full subjugation of Ukraine to Moscow (1921/2–1991) left a lasting impact on their symbolic space. The names of streets, squares, and parks as well as architecture and monuments were used as a tool to achieve a specific political goal—strengthening the image of the Soviets and maintain their domination. When the USSR was dissolved under Belovezh Accords in December 1991, its symbols in the city landscape of Ukraine were partially dismantled. The process of removing Soviet symbolism has not been uniform and has not yet been completed. This applies especially to areas dominated by people of Russian and Russian-speaking origin.

The decommunisation of Ukrainian urban space can be described in comparison to a sinusoid. When the country was ruled by the pro-Russian option (i.e. 2010–2014) or the rather flexible option (1994–2005), this process slowed down, whereas in the periods of political changes bringing Ukraine closer to the West in was intensified. There are four periods of increased dismantling of Soviet symbols (discussed in detail later):

1. directly after Ukraine gained independence after the collapse of the USSR in 1991,
2. during the Orange Revolution at the turn of 2004 and 2005,
3. after 2014 (Euromaidan, Annexation of Crimea, War in Donbas),
4. presently, as a result of a new phase of Russian aggression initiated in February 2022.

Ad. 1. Individual acts of iconoclasms of Communist monuments from the cities of Ukraine, as part of the first stage, actually appeared as early as in the 1980s in the western oblasts. Shortly after the collapse of the USSR and Ukraine's independence, the process started to spread unevenly in different parts of the country, though the dismantling of monuments to Communist leaders to the largest extent affected Eastern Galicia. The removal of the Soviet symbols from urban space was as a result of public moods related to the emergence of a new state and the collapse of the previous one, as well as hope for deeper integration with Western Europe. Similarly, in Poland, in the first period of democratisation in the early 1990s, numerous acts of iconoclasm appeared, expressing the desire of Poles to cut themselves off from the Communist past.

Ad. 2. The monuments, which survived the wave of decommunisation in the 1990s, later became the target of attacks during the Orange Revolution. Election fraud, which was intended to bring the loser Viktor Yanukovych to power, caused a surge of protests, as a result of which another group of post-Soviet symbols was devastated or overthrown. Though spontaneous, this stage was the least abundant in this type of events and ended shortly after the rightful president, i.e. Viktor Yushchenko, came to power.

Ad. 3. The substantial, full-scale decommunisation of Ukrainian urban space took place only as a result of Euromaidan (Yekelchyk 2021). In the first month of 2014, over 550 statues of Lenin were demolished across Ukraine (Plokhii 2018). This action over time started to be called 'Leninfall' or *Leninopad* in Ukrainian (Vlasenko and Ryan 2022). It is worth noting that in Ukraine there used to be over 5,500 monuments to the leader of the October Revolution, and a significant part of them survived the first two waves of decommunisation (Gaidai 2021). During the third stage, an attempt was also made to legally regulate the process of decommunising public spaces. In 2015, controversial Memory Laws also known as four 'decommunisation laws' were signed (Kozyrska 2016). They were perceived as contentious, considering that apart from decommunisation, the laws also created a potential conflict between

Ukraine and Poland by attempting to revive the memory of the 'builders' of the Ukrainian Insurgent Army (UPA) responsible for the summer of 1943 Volhynia genocide of Poles. Nevertheless, since 2015 the Institute of National Remebrance (INR) supervised the compliance with the regulations. Interestingly, the introduction of the Memory Laws coincided with the decommunisation act signed in 2016 in Poland.

It was indeed after 2014 that, for several reasons, the process of decommunising Ukrainian cities was possible to be carried out much more thoroughly. There were several factors behind this change, rightly pointed out by Oliinyk and Kuzio (2021): the Euromaidan Revolution; collapse of pro-Russian political forces; election of a large pro-European parliamentary coalition; and the impact of Russian military aggression on Ukrainian attitudes to Russia and Ukrainian national identity.

Ad. 4. The last of the listed stages of the removal of Communist monuments began after the full-scale Russian invasion of Ukraine in February 2022. The exact outcomes of this process are yet to be known; however, there are several cases of decommunisation worth enlisting. Already on 9 April, in Mukachevo in Carpathian Ruthenia, a Soviet tank, the monument to the liberation of the city from Nazi occupation by the Red Army, was removed. A few days later, a similar fate befell the monument to Soviet soldiers in Stryj (Lviv region). On 21 April, three Communist monuments were removed from Chernihiv after its liberation from the brief Russian occupation. On 26 April, in Kiev, a monument depicting the workers of two fraternal nations—a Ukrainian and a Russian—was dismantled, and so was the aeroplane-form monument to Soviet aviators in Zaporizhzhia. On 5 May, the Soviet tank monument from the Victory Square in Zhytomyr was removed, and four days later, the same happened with Pushkin statue in Ternopil. Other monuments that were dismantled after the Russian invasion are that of Nikolai Kuznetsov in Rivne, the Marshal of the Soviet Union Georgy Konstantinovich Zhukov and Alexander Nevsky in Kharkiv, as well as different statues in Drohobych, Kolomyia, Vinnytsia, and Mykolaiv. Furthermore, the wave of decommunisation of the symbolic urban space has also spread beyond Ukraine. Monuments that had memorialised the former Soviet domination of the CEECs were also removed in Poland, Estonia, Lithuania, and Latvia, what more broadly illustrates the growing reluctance of the inhabitants of these countries towards the post-Soviet, and frequently—the post-Russian legacy (see Pushkin or Nevsky) in general.

4.9 Summary and Conclusions

Within itself, a monument embodies, not merely aesthetic features, but also and first and foremost symbolic significance—and so, when iconoclasm is engaged in, this is not on account of the form assumed, but rather the content. The 1989 change in Poland's economic and political system and obvious desire to remove Communist content from the space of cities did lead to the removal of monuments erected in

Warsaw in the 1945–1989 period. By systematically reviewing incidences of iconoclasm in Warsaw in the time since Communism in Poland collapsed, as well as in Ukraine after the fall of the USRR, and by paying particular attention to the associated causes and processes (as well as obviously the content of the monuments involved), the authors of this chapter have been able to evidence the way in which these cases were associated with trends for the political change taking place in the country.

Other than in the case of the Warsaw's Monument to Felix Dzerzhinsky (whose destruction acquired revolutionary status by way of a rather chance happening), the instances of iconoclasm arising in Warsaw immediately after Communism fell were of rather a discreet nature, taking place—literally or figuratively—behind closed doors. The authorities took monuments away without attracting crowds, very often consulting or informing nobody as they did so. The activity in this case reflected the specific nature of decision-making by Poland's first democratic authorities. Thus, the policy vis-à-vis both the decommunisation of the political scene and (as we make clear) the symbolic space in Poland's capital city was one that tended to be characterised by consensus and a peaceable approach. The iconoclasm pursued in Warsaw in the first years that the Third Republic of Poland was in operation, though many in number, was not used by the authorities as a manifestation of markedly anti-Communist postulates. Instead, the process may be seen as one involving the steady removal of the most obvious cases whereby key Communist-era activists had been commemorated and celebrated, or else the previous system (*ancient regime*) lionised in an overt way.

During the time the party of post-Communists in the *SLD* was in power, no iconoclasm was engaged in Warsaw, which is also reminiscent of the Ukrainian period of post-Communist rule. That situation changed in 2005, however, when the right-of-centre *PiS* party came to power for the first time. It was then that it became possible to observe the onset of a marked polarisation of Poland's political scene—a process that has been undergoing a major further deepening and radicalisation in recent times. It has been in this period that the pursuit of unfinished decommunisation has represented one of the main election slogans for *PiS* (*inter alia* applied in the 2015 parliamentary elections), serving clearly as a means by which to accumulate political capital. The 2016 'Decommunisation Act' only reinvigorated the public debate on the need for some of Warsaw's monuments to be removed, with many controversies in this way aroused among inhabitants of the capital city. Here, also analogies to the situation in Ukraine, where similarly motivated legal regulations were adopted in 2015, can be found. Against that background, it needs to be stressed that though decisions regarding removal were indeed taken in respect of two monuments (the Monument of Gratitude to the Red Army and the statue of Gen. Zygmunt Berling), the application of the Act's provisions in Warsaw has been selective. Attesting to that are a number of Communist-era monuments remaining in the cityspace, not least the Monuments commemorating the First Army and the Kościuszko Infantry Division. The matter of iconoclasm would thus seem to be invoked in public debate to remind people that electoral pledges have been adhered to and to stress how those on the right pursuing an active policy of account-settling with the Communist past may be contrasted with the current Opposition, whose attitude to decommunsiation is presented as more moderate and muted. This is then an obvious measure seeking to placate a certain defined electorate, most especially in advance of general elections.

Warsaw's space (including that of a symbolic nature) thus represents a field in which clashes have taken place, and there are demonstrations of different attitudes to Poland's historical past to be seen, as well as divergent political views. The ultimate persisting manifestation of this is the co-occurrence in the cityspace of monuments of content relating to the former system, as well as others of a typically anti-Communist nature.

In parallel to the Polish events, the symbols of Communism have been dismantled for over three decades in Ukraine. In many respects, it is a similar process—it has a variable dynamics, it intensifies with political change, and it is still unfinished. Nevertheless, the aspect differentiating the two cases presented in this chapter is the Russian factor, that plays a much more significant role in Ukraine, a country still more influenced by the decisions taken in the Kremlin. Furthermore, it is also a country inhabited by a large national minority of Russians and by a Russian-speaking population, whose presence had an undeniable impact on the pace of the decommunisation process. In the face of the Russian invasion of February 2022, the process that was previously aimed at removing the symbols of totalitarianism from the urban space began to take the broader form of derussification, and it is treated as such by both sides of the conflict.

References

Andrusz G (2008) Berlin, Moscow, Warsaw: a century of sibling rivalry expressed in urban form. Urban Res Pract 1(2):181–198

Azaryahu M (1996) The power of commemorative street names. Environ Plan D: Soc Space 14(3):311–330

Bartetzky A (2006) Changes in the political iconography of east central European capitals after 1989 (Berlin, Warsaw, Prague, Bratislava). Int Rev Sociol 16(2):451–469

Bernhard M (1996) Civil society after the first transition: dilemmas of post-communist democratization in Poland and beyond. Communis Post-Commun 29(3):309–330

Charlesworth A (1994) Contesting places of memory: the case of Auschwitz. Environ Plann D: Soc Space 12(5):579–593

Crang M, Travlou PS (2001) The city and topologies of memory. Environ Plann D: Soc Space 19(2):161–177

Cresswell T, Hoskins G (2008) Place, persistence, and practice: evaluating historical significance at Angel Island, San Francisco, and Maxwell Street, Chicago. Ann Assoc Am Geogr 98(2):392–413

Czarnecka, D (2015) "Pomniki wdzięczności" Armii Czerwonej w Polsce Ludowej i w III Rzeczypospolitej, Instytut Pamięci Narodowej. Komisja Ścigania Zbrodni Przeciwko Narodowi Polskiemu (The Institute of National Remembrance—Commission for the Prosecution of Crimes against the Polish Nation)

Czepczyński M (2008) Cultural landscapes of post-socialist cities: representation of powers and needs. Ashgate Publishing Ltd

Czepczyński, M (2010) Interpreting post-socialist icons: from pride and hate towards disappearance and/or assimilation. Human Geograph 4(1):67–78

Diener AC, Hagen J (2013) From socialist to post-socialist cities: narrating the nation through urban space. Natl Pap 41(4):487–514

Donohoe J (2002) Dwelling with monuments. Philos Geogr 5(2):235–242
Dwyer OJ, Alderman DH (2008) Memorial landscapes: analytic questions and metaphors. GeoJournal 73(3):165–178. https://doi.org/10.1007/s10708-008-9201-5
Edensor T (1997) National identity and the politics of memory: remembering Bruce and Wallace in symbolic space. Environ Plann D: Soc Space 15(2):175–194
Foote KE, Azaryahu M (2007) Toward a geography of memory: geographical dimensions of public memory and commemoration. J Political Military Sociol 35(1):125–144
Foote KE, Tóth A, Árvay A (2000) Hungary after 1989: inscribing a new past on place. Geogr Rev 90(3):301–334
Forest B, Johnson J (2002) Unraveling the threads of history: Soviet-Era monuments and Post-Soviet national identity in Moscow. Ann Assoc Am Geogr 92(3):524–547
Forest B, Johnson J (2011) Monumental politics: regime type and public memory in post-communist states. Post-Soviet Affairs 27(3):269–288
Forest B, Johnson J (2019) Confederate monuments and the problem of forgetting. Cult Geogr 26(1):127–131
Gaidai O (2021) Leninfall in Ukraine: how did the Lenin Statues disappear? Harvard Ukrainian Stud 38(1/2):45–69
Gamboni D (2013) The destruction of art: iconoclasm and vandalism since the French Revolution. Reaktion Books
Gawlik P (2018) Pomnik Wdzięczności Żołnierzom Armii Radzieckiej przeniosą na cmentarz na Ochocie? [Will the monument of gratitude to the soldiers of the Soviet Army be removed to the Ochota cemetery?]. Gazeta Stołeczna. Available from: https://warszawa.wyborcza.pl/warszawa/7,54420,23789585,pomnik-wdziecznosci-zolnierzom-armii-radzieckiej-przeniosa-na.html. 10 Aug 2021
Gladsky TS (1986) Polish post-war historical monuments: heroic art and cultural preservation. Polish Rev 31(2/3):149–158
Grzesiuk-Olszewska I (1995) Polska rzeźba pomnikowa w latach 1945–1995. Neriton
Grzesiuk-Olszewska I (2003) Warszawska rzeźba pomnikowa. Neriton
Harvey C, Sanaei A (2015) Public monuments as loyalty signals in authoritarian states. Draft prepared APSA, pp. 1–33. Available from: https://pdfs.semanticscholar.org/95f5/13f61313885bf729427124f40d032f163357.pdf. 22 Dec 2021
Hayden D (1997) The power of place: urban landscapes as public history. MIT Press
Hoelscher S, Alderman DH (2004) Memory and place: geographies of a critical relationship. Soc Cult Geogr 5(3):347–355
Hoskins G (2004) A place to remember: scaling the walls of Angel Island Immigration Station. J Hist Geogr 30(4):685–700
Ivanov S (2009) Opportunities for developing communist heritage tourism in Bulgaria. Turizam: međunarodni znanstveno-stručni časopis 57(2):177–192
Jałowiecki B, Sekuła EA (2009) Miejskie szlaki pamięci. Stud Reg Lokalne 4(38):5–20
Jasiewicz K (2008) The (not always sweet) uses of opportunism: post-communist political parties in Poland. Communis Post-Commun 41(4):421–442
Johnson NC (1995) Cast in stone: monuments, geography, and nationalism. Environ Plann D Soc Space 13(1):51–65
Johnson NC (2003) Ireland, the great war and the geography of remembrance. Cambridge University Press
Jones P (2007) 'Idols in stone' or empty pedestals? Debating revolutionary iconoclasm in the post-Soviet transition. In: Boldrick S, Clay R (eds) Iconoclasm: contested objects, contested terms. Routledge, pp 241–260
Jones M, Jones R, Woods M, Whitehead M, Dixon D, Hannah M (2014) An introduction to political geography: space, place and politics. Routledge
Kałużna J (2018) Dekomunizacja przestrzeni publicznej w Polsce – zarys problematyki. Środkowoeuropejskie Studia Polityczne 2:157–171

Kattago S (2009) War memorials and the politics of memory: the Soviet war memorial in Tallinn. Constellations 16(1):150–166
Kattago S (2015) Written in stone: monuments and representation. In: Kattago S (ed) The Ashgate research companion to memory studies. Ashgate, pp 179–196
Killingsworth M (2010) Lustration after totalitarianism: Poland's attempt to reconcile with its Communist past. Communis Post-Commun 43(3):275–284
Koerner JL (2017) On monuments. Res Anthropol Aesthet 67(1):5–20
Kołodko GW (2009) A two-thirds of success. Poland's post-communist transformation 1989–2009. Communis Post-Commun 42(3):325–351
Kozyrska A (2016) Decommunisation of the public space in post-Euromaidan Ukraine. Polish Pol Sci YB 45:130–144
Kros C (2015) Rhodes must fall: archives and counter-archives. Critical Arts 29(1):150–165
Leach N (1999) Architecture and revolution: contemporary perspectives on Central and Eastern Europe. Psychology Press
Levinson S (2018) Written in stone: public monuments in changing societies. Duke University Press
Lewicka M (2008) Place attachment, place identity, and place memory: restoring the forgotten city past. J Environ Psychol 28(3):209–231
Light D (2004) Street names in Bucharest, 1990–1997: exploring the modern historical geographies of post-socialist change. J Hist Geogr 30(1):154–172
Lowenthal D (1961) Geography, experience, and imagination: towards a geographical epistemology. Ann Assoc Am Geogr 51(3):241–260
Mitchell K (2003) Monuments, memorials, and the politics of memory. Urban Geogr 24(5):442–459
Nelson RS, Olin M (2003) Monuments and memory, made and unmade. University of Chicago Press
Ochman E (2010) Soviet war memorials and the re-construction of national and local identities in post-communist Poland. Natl Pap 38(4):509–530
Ochman E (2013) Post-communist Poland-contested pasts and future identities. Routledge
Ochman E (2018) Spaces of nationhood and contested soviet war monuments in Poland: The Warsaw monument to the brotherhood in arms. In: Bevernage B, Wouters N (eds) The Palgrave handbook of state-sponsored history after 1945. Palgrave Macmillan, London, pp 477–493
Oliinyk A, Kuzio T (2021) The Euromaidan revolution, reforms and decommunisation in Ukraine. Eur Asia Stud 73(5):807–836
Osborne BS (2001) Landscapes, memory, monuments, and commemoration: putting identity in its place. Can Ethn Stud 33(3):39–77
Plokhii S (2018) Goodbye Lenin: a memory shift in revolutionary Ukraine. MAPA—Digital Atlas of Ukraine. https://gis.huri.harvard.edu/leninfall
Skillingstad TL (2016) Cihu Memorial Sculpture Park. Iconoclasm, memory and the importance of space. Master's thesis, UiT Norges Arktiske Universitet
Śmigiel M (2012) Czterech Śpiących wraca na Pragę. Mieszkańcy przeciw [Czterech Śpiących returns to Praga. Residents against]. Gazeta Wyborcza. Available from: https://warszawa.wyborcza.pl/warszawa/1,54420,12146455,Czterech_Spiacych_wraca_na_Prage__Mieszkancy_przeciw.html. 15 Dec 2021
Smith DJ (2008) Woe from stones': commemoration, identity politics and Estonia's 'war of monuments.' J Baltic Stud 39(4):419–430
Solarz MW (2014) The rise, fall and rebirth of Polish political geography. Geopolitics 19(3):719–739
Solarz MW (2012) The communist world from dawn till dusk. A political geography perspective. Miscellanea Geographica—Regional Stud Dev 16(1):17–22
Stańczyk E (2013) Remaking national identity: two contested monuments in post-communist Poland. Central Europe 11(2):127–142
Stryjek T (2018) The hypertrophy of polish remembrance policy after 2015: trends and outcomes. Zoon Politikon 9(1):43–66

Szpala I (2013) Zostawić czy usunąć pomnik Śpiących? Wyniki nowego sondażu [Leave or remove the Śpiących Monument? New poll results]. Gazeta Wyborcza. Available from https://warszawa.wyborcza.pl/

Vlasenko Y, Ryan BD (2022) Decommunization by design: analyzing the post-independence transformation of Soviet-Era architectural urbanism in Kyiv, Ukraine. J Urban Hist 00961442221079802, 1–42

Williams P (2008) The afterlife of communist statuary: Hungary's Szoborpark and Lithuania's Grutas Park. Forum Mod Lang Stud 44(2):185–198

Wojtczuk, M (2015) Rada Warszawy: Czterej Śpiący już nie wrócą [Warsaw Council: Czterech Śpiących will not return], Gazeta Wyborcza. Available from: https://warszawa.wyborcza.pl/warszawa/1,34862,17486935,Rada_Warszawy___Czterej_Spiacy__juz_nie_wroca.html

Yekelchyk S (2021) Symbolic plasticity and memorial environment: the afterlife of Soviet monuments in post-Soviet Kyiv. Can Slavonic Pap 63(1–2):207–228

Zaborowska MJ (2001) The height of (architectural) seduction: reading the "changes" through Stalin's Palace in Warsaw, Poland. J Archit Educ 54(4):205–217

Zaremba Ł (2018) *Obrazy wychodzą na ulice. Spory w polskiej kulturze wizualnej*, Fundacja Bęc Zmiana IKP UW, Warsaw.

Zubek R (2008) Parties, rules and government legislative control in Central Europe: the case of Poland. Communis Post-Commun 41(2):147–161

Chapter 5
Geoethics and Disaster a Geographical Approach

Dorota Rucińska

5.1 Introduction

This section combines an academic discussion of geographical issues and ethics in selected aspects. It presents what are considered to be the definitions of ethics and geoethics. Contemporary problems within this recent research field are based only on geological field, how this study shows, with results obtained from—or are combined with while finding common ground—geography.

Ethical problems concern various issues, natural hazards, and socio-economic aspects. They affect various areas and phenomena such as tax havens, poverty, and dark tourism (tourism that involves visiting places that are recent and historically associated with death and tragedy). The study discusses the legitimacy of including selected geographic issues besides geological ones related to natural hazards and natural disasters to a new field called geoethics. However, there is one special area in geoethics which focuses on the issue of natural disasters, which was a subject at the World Conferences on Disaster Reduction in Yokohama, Kobe, and Sendai.

This section reviews ethical problems linked to natural hazards and disaster, identifying the new problems in order to increase theoretical understanding of the subject. Natural disasters linked within a variety of decisions or behaviour taken at the level of government officials, as well as corporation, non-profit organizations, communities, and individuals and business actions. These decisions are crucial due to the safety of the population in risk area.

The study undertaken has a significant socio-economic significance: social due to the formation of social norms, economic due to the impact of human activities on the economy, and secondary—on society and the economy. Ethics problems directly or indirectly affect the dimension of natural disaster and ability for the Disaster

D. Rucińska (✉)
Faculty of Geography and Regional Studies, University of Warsaw, ul. Krakowskie Przedmieście 30, 00-927 Warszawa, Poland
e-mail: dmrucins@uw.edu.pl

K. Podhorodecka and T. Wites (eds.), *Global Challenges*, Springer Geography,
https://doi.org/10.1007/978-3-031-60238-2_5

Risk Reduction. Thus, it also describes the Triangle of Disaster Risk Reduction (TDRR) that includes those ethical issues raised at the conferences. The issue of corruption in this context has also been raised. This study shows that behaviours, decisions, and actions that are theoretically aimed at reducing risk sometimes go beyond ethical principles. These actions are taken at different stages, before and after natural hazards, and as a consequence might cause substantial damages and affect the population and its properties. Examples of media communications as well as politicians, public administration, resident decisions and actions that impair DRR process and increase negative effects of natural disaster are presented.

Without use the ethics in the context of natural disasters is present among both, individuals and state institutions, and the benefits of their decisions can create negative effects. Liquidation of non-ethical decisions is a challenge of DRR following the Sendai Framework for Disaster Risk Reduction (2015–2030). This study is partially a theoretical paper supporting further development of the Sendai Framework.

5.2 Ethics and Geoethics Definitions as Basis to Geographical Approach for Disaster Risk Reduction

Beginning with the definitions of ethics, we will look at the definition of geoethics indicating a geographical approach including disaster aspect.

The terms ethics and morality are closely related and are concerned with what is considered as being morally good or bad, with the term also applied to principles and ethics. It deals with questions at all levels and consists of the fundamental issues of practical decision-making. Ethical problems include the nature of what is considered to be of ultimate value and those standards of what is right or wrong by which human actions can be evaluated (Britannica 2021, www.britannica.com).

The field of ethics involves systematizing, defining, and recommending concepts of right and wrong behaviour. Normative ethics take on a more practical task, and how one should act, morally speaking. This may involve articulating those positive habits that we should use, the duties that we should follow, or the consequences of our behaviour upon others (*Internet Encyclopedia of Philosophy*, IEP).

The ancient foundation of the ethics of Socrates and Aristotle has evolved into bioethics (Schweitzer and Naish 1923; Potter 1971) and is based on the thoughts of Aldo Leopold that in turn are related to the ethics (Potter 1988). The wisdom is defined how to use the knowledge for social good and the ethics cannot be divorced from biological facts as well as limit human chances to survival in the future. According to F. Ippolito (1968), whoever possesses this knowledge is responsible in transforming this knowledge into ethical action for the common good and for public use. It was Ippolito's considerations on the ethical value of geological knowledge from which geoethics developed, now defined as a scientific discipline relating to the planet's raw materials (Nemec 1992). Geoethics is also a set of accepted rules for the use of new technologies that can affect the environment, including the people living in that environment (Treder 2006).

The International Association for Promoting Geoethics (IAPF) defines geoethics as follows: geoethics consists of research and reflection on the values which underpin appropriate behaviours and practices, wherever human activities interact with the Earth system. Geoethics deals with the ethical, social, and cultural implications of geoscience knowledge, education, research, practice, and communication (www.geoethics.org). Geoethics is also defined as 'the human-environment relationship present in sustainable development considering modern economic and social development expectations' (Bobrowsky et al. 2017). Geoethics refers to human activities interact with the Earth's natural system. It is also a tool to influence the awareness of society regarding problems related to geo-resources and geo-environment (*Definition of Geoethics,* www.Geoethics.org). According to the statement: 'who has knowledge, is ethically responsible for transforming this knowledge into ethical actions for the common good and public use' (Ippolito 1968), geoethics has developed. 'Geoethics consists of research and reflection on the values which underpin appropriate behaviours and practices, wherever human activities interact with the Earth system'. Geoethics deals with 'the ethical, social, and cultural implications of geoscience knowledge, education, research, practice, and communication, providing a point of intersection for Geosciences, Sociology, Philosophy, and Economy (…)' (www.geoethics.org/geoethics) by International Association for Promoting Geoethics (IAPF). Geoethics covers human-environment relationships that are part of the research of many geographical departments around the world dealing with the issues of the Nature-Society, including Faculty of Geography and Regional Studies at the University of Warsaw (Rucińska 2005; 2012a, 2012b).

Human-environment relationships are part of the research of many geographers and geographical departments around the world that deal with the issues of Nature-Society, and in the 1980s man-environmental relations were already included in the definition of regional geography (Dumanowski 1981; Kantowicz and Skotnicki 1984), as for example. These were the basis of both research carried out and as part of a master's theses starting from 1960 (Rucińska and Walewski 2004). Studies of human-environment relationships were also the basis for the delimitation of geographical regions (Kantowicz and Walewski 2010). One could say that the subject of human-environment relationships is spread across the entire field of geography; it is not only developed, but is still taken at various scientific centres. There has been interest in the problems observed as regards raw materials and energy in human-environmental situations in the context of a number of geological and geographical areas. These are: raw materials and energy (Vecchia 2015; Byrska-Rąpała 2008); water (Glanz and Zonn 2005; Kundzewicz and Kowalczak 2008; Limaye 2015; Datta 2015); tourism and heritage (Allan 2015; Ferrero and Magagna 2015); culture (Brilcland 1998; Peppoloni and Capua 2012; Ripp and Rodwaell 2015); dilemmas of civilization and the environment (Kantowicz and Roge-Wiśniewska 2012; Nikitina 2016); perception of natural hazards (Rucińska 2012a, b; Crescimbene et al. 2015); risk perception (Usuzawa et al. 2014; De Pascale et al. 2015, Tanner and Árvai 2018); historical memory (Segimoto et al. 2010; Rubeis et al. 2015); social awareness (Rucińska

2012a, b; Rubeis et al. 2015); education (Rucińska 2011; UNICEF 2012; Lollino et al. 2014; Silva et al. 2015; Aghaei et al. 2018). These are studies of those diverse environments, places, and spaces that are spread across the Earth's surface and their interactions.

Such an approach goes back to the considerations of determinism by Friedrich Ratzel and the possibilism initiated by Paul Vidal de la Blache. However, without an understanding of the natural world, the diversity of social relations on the planet cannot be explained. Moreover, physical and human geography servicing regional geography. But geography had entered, like many other sciences, into hyper-specialization. At the same time, geographers are demonstrating all their skills in leading very serious research on extreme events, governance of catastrophes, climate change and environmental refugees, social resilience, global tourism and global pandemic, etc. (Kesteloot and Bagnoli 2021). In parallel, they realize that the limit between what is attributable to human or society and what to the nature has become too fragility. Today, a strong divide between the human (represented by human geography) and the physical (by physical geography) approaches is no more advisable. Here comes the perspective of regional geography (Dumanowski 1981; Kantowicz and Skotnicki 1984; Rucińska and Walewski 2004) currently perceived as a 'new regional geography' based on human and physical relationships (Kesteloot and Bagnoli 2021). Finally, understand that the limit between what is attributable to human or society geography and what to natural environment or historical geography has become too weak. Therefore, a strong divide between these approaches is no more advisable. Finally, we understand that the limit between what is attributable to human or society geography and what to natural environment or historical geography has become too weak. Therefore a strong divide between them approaches is no more advisable. The problem is on the verge of sub-disciplines and physical geography, regional geography, the ethics, climate change aspects, and the Disaster Risk Reduction (in risk management). Thus, we address the really topical issue in the scientific discipline of geography (Cornut and Swyngedouw 2000; Storper 2011; Whatmore 2014; Asheim 2020; Taylor and O'Keefe 2021; Stryjakiewicz 2022; Martin and Sunley 2022; Weckroth and Ala-Mantila 2022). Time to regionalize the world in accordance with the local effects of global problems (such as climate change and natural hazards) and local opportunities to solve these global problems, including societies in Disaster Risk Reduction, and respecting human-nature relationships.

Many authors also refer to ethics in the context of fields and scientific issues. For example, the problem of differentiation as regards the quality of socio-economic life in an era of globalization (Borowiec 2006); business ethics in the context of sustainable development (Kuzior 2016); and geoethics (Peppoloni 2012; Peppoloni, Di Caoua 2012, 2015, 2017, Wyss and Peppoloni 2014; Rucińska 2016, 2017).

This part is set out to review the geoethics problems as regards taking geographical approaches and attempts to integrate the geoethics concept. It presents some of the problems and focuses on the concept of natural Disaster Risk Reduction in order to increase our theoretical understanding of the subject. This paper also presents the Triangle of Disaster Risk Reduction (TDRR) that shows the main elements of natural Disaster Risk Reduction.

The chapter is a geographical perspective to geoethics as well as the beginning of a geographical discussion to understand the role of widely understood geoethics in the Disaster Risk Reduction (DRR).

There is a long academic tradition in discussing ethics and ethical theories. The field of ethics involves systematizing, defines, and recommending concepts of right or wrong behaviour. Normative ethics takes on a more practical task, which is to arrive at moral standards that regulate right and wrong conduct. This may involve articulating the good habits that we should use, the duties that we should follow, or the consequences of our behaviour on others (*Internet Encyclopaedia of Philosophy*, IEP).

Ethics should influence people's morality. The challenge of normative ethics is to answer the following questions: how to use the rules in practice; how to relate them to general principles, behaviours and decisions; and what are the moral standards that regulate good and bad actions and attitudes (Gryżyna 2009). In the context of natural hazards and decision-taking, there are three elements that influence actions of man:

1. Perception as a structure of information about natural hazards can be obtained both sensually and verbally (Pocock 1974) and are important in undertaking preventive, rescue, and compensatory actions (White 1974).
2. The role of awareness is marked by many researchers; the awareness is a factor that affects on quality of decisions, preparedness, and protection some self, too (Kates 1971); before, during, and after. The awareness allows us predict effects of natural hazards.
3. Attitude: Reaction of the people on natural hazards can be different, as well as factors of warnings ignore (Palm 1981).

Higher awareness influences the ability to undertake protective actions, including individual ones (Kates 1971), and plays an important role in minimizing the effects of natural hazards (Hanson et al. 1979).

Perception and knowledge about the risks and their effects play a role in shaping the attitude of the human being as a consequence, leading to specific behaviours and decisions. This figure is complete with ethical principles (Fig. 5.1).

Therefore, it is important to put ethics in the triangle of Disaster Risk Reduction. The 'Triangle of Disaster Risks Reduction' (TDRR) presents the society as an important link that receives information and warnings about the imminent threat, as well as those taking part in preventive activities. The TDRR includes: (a) hazard and risk assessment and maps creation, (b) systems' creation and modification of the information, warning, risk and DRR management, prevention, (c) respond including the society, in three phases: before, during, and after the defeat (where after passes flows in before); based on knowledge and social education (formal and informal). These actions require observance of moral principles. Therefore, ethical elements falling within the scope of geoethics should be included in them (Rucińska 2014).

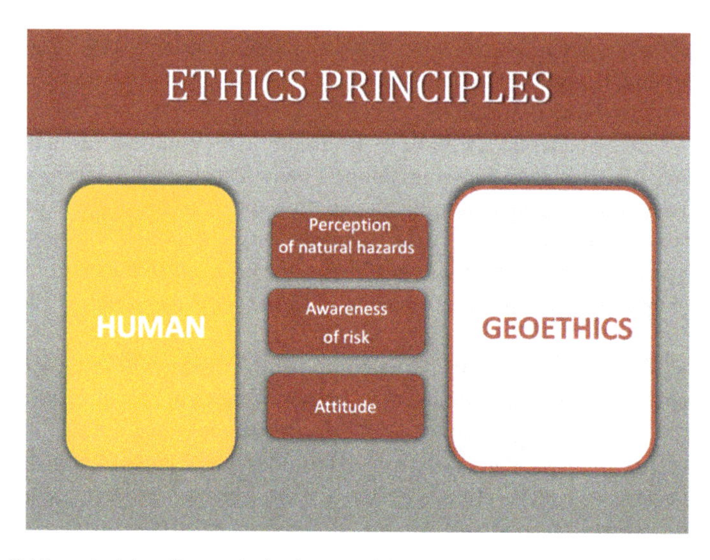

Fig. 5.1 Ethics principles. *Source* Author's own elaboration

5.3 Acceptance of Disaster Risk and Moral Hazards' Phenomenon

Disaster Risk Reduction is influenced by disaster risk acceptance (and also natural hazards acceptance, for example, dark tourism). Ethical problems are faced by people forced to make individual decisions in the face of threats or as a result of decisions resulting from the performed function. Making a choice depends on the level of risk acceptability, which is an individual feature or results from professional procedures or the lack thereof. The topic of risk acceptability in the context of natural hazards has not yet been thoroughly researched and developed. However, it can be said that acceptability is related to the characteristics of a given culture, as well as individual characteristics, including empathy. It can be assumed that each person may have a different level (threshold) of risk acceptability, depending on the kind of natural hazard. On the other hand, every person consciously or less consciously uses his own risk assessment every day and decides whether to exceed (or not exceed) the threshold impacting the consequences, profit or loss. In the context of the risk of a natural disaster, the acceptability of the risk can be considered at three social levels of human activity (individual, community or group, and institutional or business) (Fig. 5.2).

Individuals may perceive and assess insecurity differently, partly due to a subjective assessment of the natural hazards or risk, which affects the level of risk acceptability. However, it should be remembered that the acceptability of risk may vary depending on the type of natural hazards and disaster and depending on the entity (i.e., in the case of society, we must take into account a large variation in the level of risk acceptance and the lack of a sense of obligation to take any action due to social

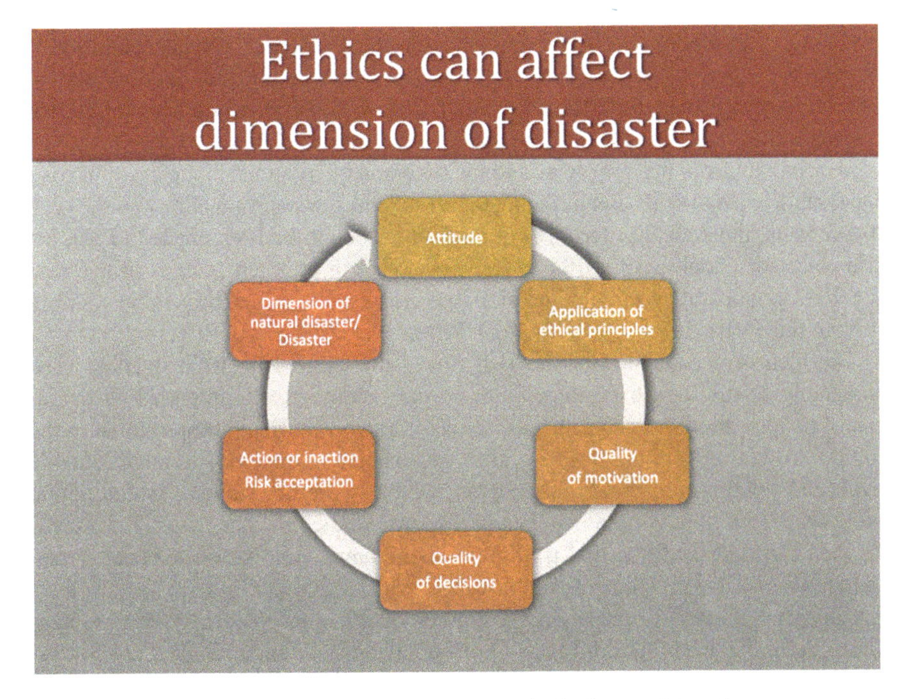

Fig. 5.2 Ethics can affect dimension of disaster. *Source* Author's own elaboration

vulnerability or moral hazard). The high level of acceptability is well illustrated by individual windsurfing during a storm in the Baltic Sea or group climbing—unacceptable for other people; or charity help in times of crisis, or visiting natural disaster sites (dark tourism).

People generally accept a higher level of risk for activities they actively choose (i.e., driving their own car) than for activities in which they are passive participants (their participation in the risk is unintentional (i.e., when driving as a tram passenger); hence it stands out voluntary (intentional) risk and unintended risk. They also accept a higher level of risk related to natural hazards than to hazards resulting from human activity (they more easily accept a flat of choice in floodplains than unintentional living below a dam or air pollution) (Grundy 2008).

We also enter the ethical sphere by establishing theoretical criteria for a natural disaster, such as: criteria for the acceptability of the number of people killed or affected, or by determining the decision and timing of delivering warnings to the public or information about evacuation (when the level of determination is determined by the institution). The acceptability of risk is also revealed in some architects who, due to the rarity of the risk greater than specified in the construction rules, do not take extreme phenomena into account, ignoring the potential social effects and loss of human life in their projects (Walker et al. 2011). Officials do not have the necessary knowledge to make decisions about the number of evacuation sites and the choice of vertical-evacuation (TVE) tsunami sites or escape routes. What is needed is a spatial analysis model that takes into account ethical criteria. The acceptability of the risk is related to the designation of escape routes taking into account ethical criteria. Escape routes, their location, and patency should take into account ethical criteria. In the case of roads for cars, such a criterion is the ability to accept a given number of cars from nearby towns and the adoption of solutions in the event of road impassability (road pollution or damage, traffic jams during evacuation); in the case of footpaths, the ethical criterion is to take into account the time needed to reach a safe place for people such as the elderly, disabled, or children (i.e., as in the case of a vertical tsunami evacuation location; the greatest distance to hills in the vicinity or from the TVE) (Engstfeld et al. 2010, Wood et al. 2014).

The idea of acceptable disaster risk can also be based on what is considered acceptable across industries, depending on the voluntary or unintended participation of people. Moreover, the acceptable risk is always inversely proportional to the number of fatalities. In the business, there is a need to reduce risk to an acceptable level, and for this purpose, risk acceptance criteria are determined, which define when specific actions should be taken in response to the risk that arises. Two stages are then distinguished: the first is without taking any action (risk acceptance state) and the second is with taking action (the state without accepting the risk). It follows that an acceptable risk is one when 'the amount of risk is acceptable for the organization without the need for additional actions or changes in functioning' (Wróblewski 2014). Continuing the considerations, risk acceptance is a conscious decision about the organization's readiness to expose itself to a given risk (ISO Guide), while the acceptable level of risk is a contractual value and the result of risk assessment based on the adopted criteria (whether the expected risk is within the acceptance or tolerance limits). The risk is within the tolerance limits, but beyond the acceptable level, should initiate risk monitoring, control, and reduction activities (Wróblewski 2014). Crossing a certain limit is said to occur when a social reaction occurs as a result of exceeding the general limit of social acceptance for risk (Rucińska 2021).

In some sense, risk acceptance criteria distinguish between 'good' and 'bad' systems and activities with regards to the risk they expose the society or elements of a society to and there is thus an obvious link between ethics and risk acceptance criteria or to risk management at large. Risk acceptance criteria, which are normative statements of what is deemed acceptable and what is not in a society, are intimately related to ethics, ethical theories, and ethical values and norms. Risk acceptance criteria have distinguishing between levels of risks that are acceptable ('good') and levels that are intolerable ('bad') describing a systems and activities with regards to the risk they expose the society (Vanem 2012). This exposure to natural hazard is link to the risk management. Therefore, in management, the criteria themselves should be based on ethical considerations and be justifiable by some ethical theory and the various principles and the philosophies may complement each other in order to achieve the overall safety objectives of society. In particular, this is useful when criteria are subject to debate and should be communicated and justified towards the public (Vanem 2012).

Another major issue, mentioned above, is the phenomenon of moral hazard, which relates to responsibility. The term moral hazard dates back to the seventeenth century and was widespread among English insurance companies from the end of the nineteenth century. Early use of the term carries negative connotations, involving embezzlement or other immoral behaviour—usually of the insured. Eighteenth-century mathematicians who studied decision-making processes used the term 'moral' as 'subjective', which may cast a shadow on ethical connotations (Dembe and Boden 2000) and separateness from ethics. Yet the research conducted by economists in the 1960s by (Pauly 1968) did not refer directly to fraud or immoral behaviour. The discussion on moral hazards continues (Pauly 1968; Baker 1996; Winter 2000; Rowell and Connelly 2012) and the term is used in many areas (i.e., insurance, health, economic crisis, trade, bank management, and digital economy). As Baker (1996) has pointed out *'(...)'moral hazard' is one of the most important, and least well understood, of the analytical tools applied to these and other social responsibility questions. Whether the topic is products liability law, workers' compensation, welfare, health care, banking regulation, bankrupt (...) law, takings law, or business law, moral hazard is a central part of the law and economics explanation of how things as they are came to be'*. Moral hazard has many definitions (i.e., economic, ex ante behaviour *'impact of insurance on the incentives to reduce risk'*) (Winter 2000); microeconomics *'lack of incentive to take care is called moral hazard'* (Varian 2010); *ex post* moral hazard concerns the effects of incentives on claiming actual losses (Abbring et al. 2007, p. 1); *'In the case of the insurer having professional skill (i.e., has received relevant news by mail) or experience having with natural phenomena (i.e., he is aware that the weather will be good), who knows that the real value venture risks are less than what the current market place estimates and who does not reveal it to the other party, still the market price is to be considered just, because probable profits derive from his professional ability'* (Lessius 1605, as quoted in Ceccarelli 2001) (as quoted in Rowell, Connelly 2012). Recently, it is a typical problem of modern economic system. *'Moral hazard, the risk one party incurs when dependent on the moral behaviour of others. The risk increases when there is no effective way to control that behaviour. Moral hazard arises when two or more parties form an agreement or contractual relationship and the arrangement itself provides the incentive for misbehavior by insuring one party against responsibility'* (www.britannica.com). An actor who has an information advantage begins to strive to maximize his own benefits at the expense of the other party who does not know this.

In the theory of economics, moral hazards are the propensity and tendency to take excessive risks because of the awareness of actual costs being borne by parties other than those involved in the hazard. This leads to economic inefficiency. Originally, this theory concerned usually insurance frauds, but later it referred to inefficiencies (and not ethical rules) that may arise when risk is transferred from one entity to another. In the insurance market, moral hazard refers to a situation where the existence of an insurance contract changes the behaviour of the insured, starting from the moment he ceases to bear the full costs of the consequences of his proceedings. Thus, moral hazard means a change in the behaviour of a person who, i.e., after taking out insurance, ceases to be strongly motivated to take actions to prevent the occurrence of a loss, which increases the likelihood of adverse effects of the insured event. For example, a person getting car insurance may be less careful in icy roads or in a storm, or start driving more than before in such weather, increasing the risk of an accident. This behaviour is due to the fact that the risk is almost completely transferred to the insurance company. The problem of moral hazard is also well known among insurers and academia (Baker 1996)—shifting many burdens and liability for performance of some activities on an individual employee, despite the fact that the institute and university are assessed as a whole.

The phenomenon of moral hazard may be feared in various ways, due to the citizen's lack of a sense of duty and to take any action in connection with preparation for natural hazards (natural disasters) and activities aimed at reducing their socio-economic vulnerability to natural hazards (usually in the case of a citizen living in welfare state). The reason may be 'excessive' compensation of losses by the state authorities, which leads to a decrease in individual mobilization of citizens at risk. This is a typical phenomenon typical of post-communist countries. With medical insurance, risky behaviours may occur related to taking up activities that directly threaten health, such as extreme sports (risk acceptance). After signing the contract, as in other cases, most of the potential costs—i.e., treatment—are passed on to the insurer. Another example is private clinics treating relatively young patients, usually healthy people, not burdened with a group of difficult-to-cure diseases, avoiding treatment of people who do not predict a relatively quick recovery. Treatment contracts offered by commercial clinics are usually signed until retirement age.

In order to reduce the moral hazard, insurance contracts use various conditions and exclusions that are intended to induce the insured to behave in a manner that limits the risk of loss. For example, the so-called the deductible or excess oblige the insured to cover part of the damage—a certain amount of money is deducted from the claim amount. These clauses are designed to eliminate small claims. The principle of co-insurance is also applied—it is a percentage share in the coverage of the damage by the insured and the insurer. Agreements may also include responsibilities.

The moral hazards' phenomenon (on the one hand, moral hazard, on the other hand, reduced sense of responsibility as a result of compensation, passive waiting for a financial reimbursement of the damage caused) causes various effects and makes it difficult to undertake local adaptation activities of the community (including educational, because the insured do not see the need to participate in in education) to adapt to natural hazards and reduce the effects. This phenomenon also affects the efficiency of management and relations: government—insurance corporation—local government and commune—information transfer—citizen.

The catastrophe is accompanied by the phenomenon of moral hazards, a strong temptation to use the catastrophe for financial and political gain, which increases existing inequalities. The 2004 tsunami destroyed many documents, including land tenure, cutting off landowners' access to land in Leam Pom. Disasters lead to environmental degradation, and the restoration of, for example, wooden houses and boats, typical of Thailand, exacerbates (often illegal) logging in Aceh (Walker 2005).

It is a Contingent matter whether certain goods and services can best be provided, at any given time and place, by government employees or those paid by private firms (Zack 2009).

The unfair terms and conditions of insurance are also noticed (a dishonest undertaking not applying the principle of not causing customer suffering). This applies to the example of a situation in Poland, where it is difficult to determine, for example, a storm or the passage of a whirlwind breaking the roof of the house (it is required to prove by the injured person, using an official confirmation issued by meteorological institutions, the passage of a storm or a whirlwind with a given wind speed; simultaneous lack of an exact coverage of the country with a network monitoring such weather events locally); or failure of such problems to be resolved by government officials will lead to non-refunding of money.

5.4 Identifying Geoethics Problems Within Contemporary World

Historically, large hydrological and transport constructions were often characterized by a high mortality rate of employees and as such are now difficult to accept from an ethical point of view. Of particular note was the period of the end of the nineteenth century, where 120,000 workers working on the Suez Canal—which was completed in 1869—died of various classifiable diseases (Ogen 2008). The Panama Canal, completed in 1914, resulted in nearly 28,000 deaths (McCullough 2001); the Three Gorges Dam has resulted in the loss of 100 employees' lives (www.interestingengineering.com); and in the U.S.A., the Hoover Dam, where, as a result of various incidents, 96 people (www.usbr.gov) died. Hundreds people also died during the construction of the Transcontinental Railroad in the U.S.A., although these were referred to as voluntary employees.

But then there those secret numbers of deaths in forced labour camps—those victims such as the prisoners of the USSR gulags. There was the construction of a road known as the Kolyma Route, built in the 1940s, where the dead would be buried in the concrete foundations of the road, and hence given the name 'Road of Bones' (Wites 2008). 3.004 prisoners were dead during the construction of the Route (www.gulag.info.pl). However, the number of victims is unknown in the case of the construction of the Transpolar Railway Bus, in Siberia along a line of latitude close to the Arctic Circle. After the outbreak of the Cold War, the Soviet dictatorship began building railway tracks with a length of nearly 8000 km through permafrost tundra (the Salekhard-Igarka Railway). The construction was interrupted shortly after Stalin's death (1953), but between 1949 and 1953, 1.500 kilometres of railway track had been laid (Mausolf 2011).

Human activities generate ethics problems that relate to man-nature relations as well as specific relationships between people in the face of natural phenomena, which can be seen in many issues concerning duality, the function of political boundaries (Kałuski 2017) and natural resources (Foltz 2002). They can be seen in many issues concerning duality, the function of political boundaries (sometimes overlapping physical and anthropogenic boundaries), and their role as a spatial barrier as described by S. (Kałuski 2017), noting that one feature of such a border should be the limits that meet ethical conditions. You can include—for example—tax havens, globalization, placing state borders on maps as a result of annexation, using the terms Third World, Fourth World, Poor South, prostitution in cross-border areas, migration, limited access to holy places, delimitation of the cosmos (no boundaries in space), and movement of pollution across state borders (Kaluski 2017).

Some of them are linked to natural disaster management, environmental migration, dark tourism, and corruption, urbanization, spatial planning, building, and tax havens.

The existence of tax havens allows us, on the one hand, a wider opportunity to obtain financial savings for companies and business development. On the other, these divert capital from the country that generates this capital and consequently deepens the differences in social wealth and so strengthens poverty. Globalization facilitates access to information and goods and opens the door to investment in various regions of the world; it facilitates international cooperation in, for example, the protection of human rights and the environment. It also facilitates the use of international goods and culture, but conversely leads to a gradual disappearance of tradition and a reduction in the diversity of cultures. Then there is the cartographic, political, and physical aspect—'soft legalization' of the state's informal borders that arise as a result of the annexation of territory of another country, which diminishes the *de facto* political significance that existed during aggression and annexation of the borders of the invaded country.

Other effects of globalization include grading a region's affluence through nomenclature, depending on their development and economic resources using deprecating names or indicating a lack of moral sensitivity as for example: Second World countries (socialist countries), the Third World (here, defined as the so-called developing countries), the Fourth World (the poorest groups of these countries), and the Poor South, which suggest a lower social class (affluence) on a global scale.

The ethical problem is the use of geometrical boundaries without taking into account cultural phenomena. This includes either excessive availability or even restriction of tourism to tourist facilities for political or religious reasons. Then there is prostitution in cross-border areas where there is usually ethnic and cultural diversity that, on one side, increases the interest in these types of services, but on the other makes it easier to hide such practices.

State borders that constitute the basic guarantees of security for their citizens are at the same time a barrier for illegal immigrants. Sometimes there is unofficial consent to their inflow; while this can bring hidden benefits to the state in the form of cheap labour, it is also an opportunity for exploitation and often superhuman effort of immigrants. Reasonable ethical questions arise from the delimitation of the cosmos (for example, the placing of satellites) and the definition of fair principles as regards its use for suborbital flights for scientific missions and/or for human transportation and observation; what determines the location of a satellite or who has the right to observe individuals from space; and who can observe the environment—along with people using drones for such purposes, and when.

As regards outer space pollution, who is obliged to clear the space junk orbiting around the globe itself? In addition, how should we deal with pollution that moves away from its geographical locational source and in turn affects those individuals who have no influence on how such pollution is generated, along with the general emissions and their presence in their geographical space? (Examples include smog, odours, and pollution from animal husbandry or industrial production, air pollution with radioactive compounds, water pollution, and water and air pollution because military conflict). How does one proceed in the vicinity of those reactor locations near the border with another country, thus exposing those individuals who are not actually benefiting from goods produced such as atomic energy, or not even agreeing to having them placed in such a nearby location? What about the ethical problems of diverting water (or wastewater) resources to other areas in methods that the natural hydrological system could manage, but in a way that disrupts this system in a country or region located in the lower regions of the river, and so reducing the standard of living and development opportunities in those areas?

These problems particularly occur when the spatial location of industrial waste or harmful substances is changed and results in natural disruption outside of the manufacturing region, even if there are financial benefits. For example, shale gas horizontal fracturing (also known as fracking) at state borders could be a problem, as this action would violate those borders. Similar situation is drilling close or under protected areas.

The real ethical problem is the various types of natural hazards and disasters as well as catastrophes that arise as a result of exceptional weather, ocean, and geological conditions. These disasters are sometimes a result of human expansion and interference in the environment that generate such problems, two examples of duality—those situations in which two opposite ideas or feelings exist at the same time. Depending on the spatial extent, we should also mention large buildings and activities in outer space, as well as natural hazards and natural disasters.

Building dams and so-called large buildings is not a new issue, and the last building of such a structure was the Three Gorges Dam in China (completed in 2010). The dam is intended to stimulate economic development (of a long-term nature), improvement of economic conditions (short-term: employment during construction, long-term: energy, security, water resource gathering, protection against floods, and offsetting its effects). However, in addition to these benefits, there are also controversial effects in the form of compulsory displacement of the population (1.3 million) and the breaking of local ties and traditions, along with permanent change to the landscape. It has also created new hazards in the form of tectonic movements that threaten breakage of the dam, and flooding in the areas of urban development or urbanization, that is, a situation in which it will become necessary to widen the buffer zone and resettle the inhabitants for their protection.

In many cases, it is difficult to pinpoint precise data, and sometimes data that is even unreliable. Changes brought about the end of slavery, the Cold War, and the development of technological solutions and workers' rights. Today, such large-scale construction no longer takes such a toll on so many accident victims; in the past, unions were often the cause of forced displacement of sometimes thousands of people, sometimes hundreds of thousands of people. In the case of the Three

Gorges Dam, however, construction still poses a threat to the inhabitants despite the casualties, even though this was supposed to protect people from flooding.

Any doubts about the implementation of these structures would require discussion of geoethics as part of DRR. This problem is related to sustainable development. Sustainable development has had considerable attention devoted to it in recent years, with one important issue being the enabling of such development while maintaining the values of the natural environment. It is also important to preserve and create good living conditions for people in relation to what is considered to be the concept of 'quality of life'. However, there has been a deterioration in living conditions, and this is not only due to the extraction of natural resources, commercial failures, or accidents. For example, the construction of windmills for the production of renewable energy in an area close to small settlements—especially farms and isolated houses—causes a shadow flicker effect that has significantly reduced the quality and comforts of life. The same results apply to gas drilling rigs, with their 24-hour a day construction followed by the extraction of this natural resource. Due to night-time operations and platform lighting, light pollution is often prevalent and so can affect nearby homes. In the case of shale gas exploitation, work is not only periodic, but intensive; often dirt tracks that run through forests and fields are the only means of transporting the gas by large trucks to and from the works. As a consequence, transport is often churning up substantial amounts of dust that drifts close to housing located nearby. Even if there are just a few of these homes or habitats, this does not mean that the problem is not there or should be ignored. When it comes to establishing a location for such industry, what is determined to be a reasonable distance from homes by experts is not usually done in consultation with local residents.

What sustainable development is aiming at is to maintain the balance and security of the natural environment and human activity. Sustainable development was also mentioned during the World Conference on Disaster Reduction in Kobe, and the human security aspect should be considered in the context of those natural disasters resulting from natural phenomena and human presence in those risk areas.

The sustainable development is an aspect of DRR. An estimation of social vulnerability to multi-hazards within the sustainability development may mitigate the disaster risk.

5.5 Dualism. Moral Sensitivity and Indifference in Mass Media, Tourism, Decision-Makers, and Business Stages of Interpretation of the Situation

A motto by Zack (2009) (below) is next part of this chapter presented some aspects of disaster ethics referring also to dualism. There are mass media, kind of tourism, administrative decision and management of risk, as well as a business approach and perspective to hazards.

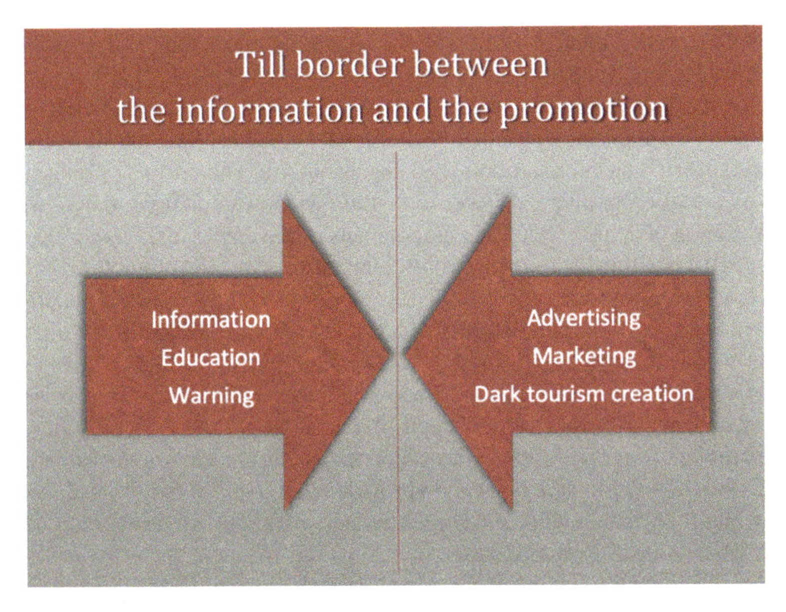

Fig. 5.3 Till border between the information and the promotion. *Source* Author's own elaboration

'Disasters also generate narratives and media representations of the heroism, failures, and losses of those who are affected and respond'

In the event of natural disasters, the mass media plays an important role, especially television (information, education, warnings) (Fig. 5.3) in transmitting information, but also, in order to interest the viewer in a broadcast or transmission; it uses more and more unusual images and comments. Some journalists even put their health at risk to create the best report or scene report.

Such cases are observed in various countries. Exciting films become the actual seed of interest in the place among certain groups of viewers. Indications present the direction of departure to people interested in dark tourism (Rucińska 2016; Min et al. 2020). More precisely, it can be called natural hazards and disaster tourism—tourism of natural hazards and natural disasters (Rucińska and Lechowicz 2014). It is compared to the former watching of gladiatorial fights in Rome (Stone 2006) and in London during the period of Roman Britain between 120 and 160 AD. This tourism includes an element of the current dynamics of the phenomenon and its drastic consequences—shortly after the event. At the same time, it is a demanding type of tourism and is classified as soft, alternative, and qualified tourism (Rucińska and Lechowicz 2014). It differs from the positively perceived thanatotourism (Rucińska 2016), defined as the tourist's interest in death in terms of the place and memory of human tragedy in the historical, educational, and cultural dimensions (Tanaś 2006).

Media activities sometimes have contradictory functions, because they not only inform and educate, but also inspire by setting the destination of the trip, and inspiration may come from the way of presenting the information material (they warn against the threat of some viewers and are deterrent for them, while among other viewers they arouse interest and emotions triggering adrenaline and provoking to face the phenomenon and touch it directly) (Fig. 5.4). An example of this is numerous reports on weather conditions, where journalists struggle with strong winds during a hurricane, fall into the water due to the impact of a sea wave on the shore, tell about the situation on debris, rubble, and damage or fire. These types of images from the scene are also commercial material that allows you to increase the viewership of a TV station or achieve flattering assessments of the employer or even keep a job (survive in the position in the era of strong competition), specific promotion of people, TV station or news programme, as well as places of natural disaster, generating a new phenomenon of tourism natural hazards and natural disasters. Sometimes, they provoke to track the threat during its duration or shortly after its occurrence (the route of a whirlwind or tornado, the place of a flood, a hurricane strike, a volcanic eruption, and an earthquake or the effects of a tsunami.

The problem of professional ethics of journalists has recently returned in the era of the COVID-19 threat and the need to provide reliable information (Hadžialić and Phuong 2020). Wuhan in China becomes the new dark tourism destination because of COVID-19 (*The* Telegraph 2020).

Focusing on natural hazard and disaster, various types of dualism can be distinguished, from dualism in geographical terms (dualism of benefits and harms in the

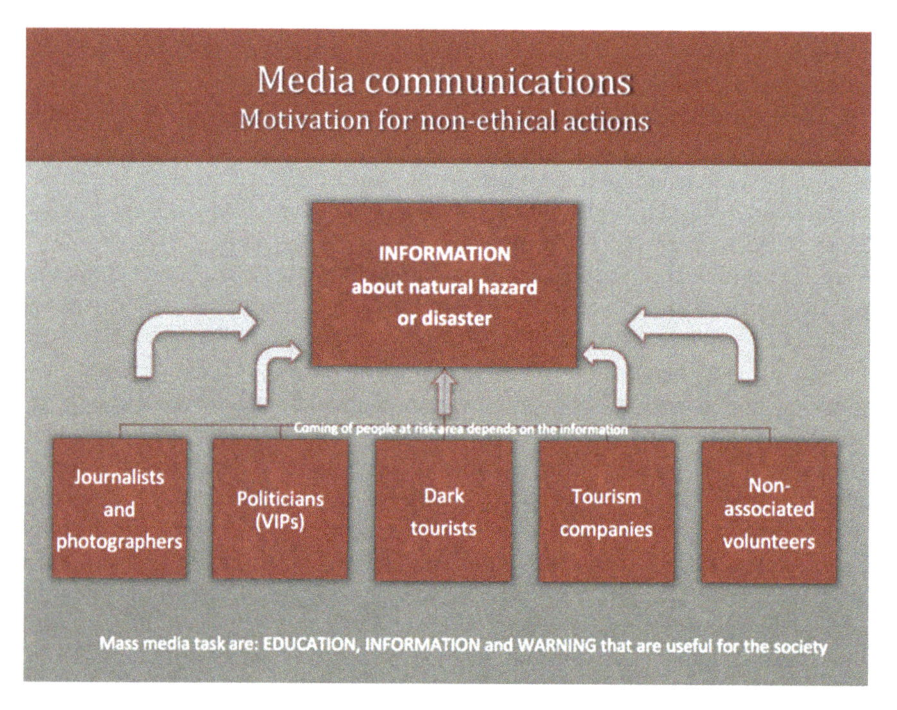

Fig. 5.4 Mass media motivation for non-ethical actions. *Source* Author's own elaboration

natural—human relationship) to dualism in philosophical terms. Behaviour and interpersonal relationships caused by a natural disaster also perceive a dualism: principles—associated with the manifestations of vulnerable and indifference to the background of benefits and harm; ethical—associated with moral choices and bad choices; political—related to the motivation of action about contradictory situations; and epistemological and logical based on contradictory judgments.

Natural hazards and disasters are related to the dualism paradigm and generate conflicting attitudes, behaviours (actions), and motivations. In terms of geography, they pose a threat to humans (natural hazards), but also generate favourable conditions of the natural environment that favour agricultural cultivation or the presence of natural resources. A natural disaster is a state of destruction and disorganization, but also a situation where, after destruction, in a relatively short time, it is possible to implement modern technological solutions limiting losses in the event of further extreme natural phenomena occurring in the future. The idea of Build Back Better also applies to this—using the time of reconstruction to create qualitatively better, more modern solutions.

Dualism is also present in the context of the economic importance of dark tourism, which causes, after a short (up to 2 years) reduction in interest in the destination, an increase in the arrival of tourists and may help in reconstruction (Rucińska and Lechowicz 2014) and accelerate the return of economic stability after the earthquake. Research in Nepal after a very strong earthquake in 2015 (the magnitude 7.8 and aftershocks). The reduction in business and group arrivals is accompanied by holiday, travel, and religious activities, as well as visits to places resistant to disaster (with a resilient tourist infrastructure). The extent to which a disaster affects tourism and the economy is subject to significant fluctuations, which may depend on the nature of the place visited (Min et al. 2020).

In the emotional sphere, both people's sensitivity and indifference are observed at various levels of functions and social roles. Sensitivity and indifference are the factors of the motivation of actions (or the lack of them) that determine the quality of the choice of behaviour (egotistical or altruistic).

The lack of moral sensitivity to other people proves the immaturity of a person, i.e., the inability to see the moral dimension of life and respond adequately to it. 'Moranne blindness', according to D. von Hildebrand, will manifest itself in focusing on one's own satisfaction, called by Zygmunt Fredu 'pleasure'. Thus, the dimension of this sensitivity is the ability to face reality, and sensitivity is the stage of mental development. Nowadays, there is a divergence between the ethics of rules and situational ethics, and indifference becomes a component of people's attitude to the world, public affairs, and human situation, i.e., in a state of weakening the sense of ethical duty towards others and the pursuit of individual goals (Żardecka 2017).

By observing the reality, in the case of decisions made by public officials, we can state moral sensitivity (presence at the site of a disaster, donation of funds for reconstruction) and indifference (not developing or not implementing solutions to problems), both of them resulting from motivations, such as willingness to help and improve the image and maintaining the position.

Referring to the ancient tradition and common sense at the same time, it should be taken into account that actions are the result of something deeper, i.e., personality (anti-behaviourist position). In the context of natural disasters, one should consider whether there are people with the appropriate personality in the decision-making positions about the fate of people affected by a disaster, and ethics is its element?

Duality is also noticeable in the mass media. The transfer of information and dedication resulting from taking up the issue helps the local community. However, many journalists' actions suggest, as in the case of politicians, and sometimes artists, taking care of their own interests and the image of a leader or representative. At the same time, a good commercial photo or video (good material) has commercial value, where the recording of unusual situations, pain, mental state, or damage is used for its own purposes. Such behaviour may be treated as indifference; however, the same photos used by the author are used by affected to defend their rights in contacts with local authorities.

Moral sensitivity and indifference (and even social insensitivity) initiate new socio-economic phenomena, such as tourism to areas of natural hazards and natural disasters. Specific examples in this case are developed countries which are proud of good education and extensive access to information. They take part in creating this tourism, which is accompanied by indifference to the current situation or the desire to satisfy one's own emotional needs. Direct observation of people at the scene of a tragedy raises reservations about moral standards. Indifference may be demonstrated by the arrival of individual tourists to places freshly mourned, not cleaned up and not rebuilt, with the community in shock or trauma after the catastrophe. Similarly, it is during organizing trips to such places (called educational excursions), because help in the form of money left behind is paid for with additional suffering by the affected. Depending on the motivation of the trip to the place of the natural disaster—whether it is to meet one's own needs or help the local population, the moral assessment will be extremely different. Likewise, the motivations of the tour operator, such as volturism, are important. However, it should be remembered that these actions could serve to defend dignity and improve the extremely dramatic situation among the affected.

The sensitivity (moral sensitivity) is defined, among others, by as the ability to feel sensory, aesthetic or ethical impressions, i.e., strong impressions, deep feelings and emotions (related to affectivity), but also to act under their influence (there is no place for passivity with sensitivity). Sensitivity is also defined as the ability to perceive the moral aspects of reality (Żardecka 2017). Others also distinguish between social and political sensitivities (Oxford Dictionary of Philosophy 2021). The sensitivity to human suffering related to the loss of loved ones, loss of health, and the possibility of normal functioning is evidenced by charity concerts organized by musicians, e.g. Hope for Haiti Now: A Global Benefit for Earthquake Relief broadcast by studios in New York, Los Angeles, London (in January 2010), as well as several dozen Polish artists Jazzmani for New Orleans—Solidarnia in Music for the victims of Hurricane Katrina, realized in Warsaw (September 2005). Non-profit organizations are also active, such as Music for Relief, an example of which was the recorded album 'for Haiti', among others, in cooperation with foundations such as:

UN Foundation and Habitat for Humanity. Moreover, in the month of the earthquake in Haiti in 2010, many institutions officially declared their help: in total, in the case of the European Union, declarations reached EUR 430 million (www.wiadomosci.wp.pl) and the World Bank—USD 100 million (www.web.worldbank.org) for reconstruction, which was widely reported in the mass media. In addition, pledged aid from the USA amounted to USD 37 million, and politicians offered civil, military, and humanitarian aid (medical care came from Cuba—startribune.com). Another expression of sensitivity was the five-year help of the Red Cross and the International Federation of Red Cross and Red Crescent Societies.

Local residents and neighbours, who are partly involved in a threatening situation due to their presence in a given place, provide heroic help. For example, in Poland we have repeatedly observed the attitude of people's sacrifice and altruism in the face of floods; the local community sparing their own hands, private cars to transport people and sacks with a belt, creating a dam with sandbags, putting their own life and health at risk. The question is, are these examples of heroism necessary for managing the crisis and the risk of a natural disaster? It seems that the share of leadership should be kept to a minimum, and management decisions should not assume the use of spontaneous help from the population in the face of a natural disaster.

This is related to responsible (in line with the principles of ethics: who knows who is responsible) risk management and conducting (or not) a well-thought-out preventive policy and investment in its implementation, based on diverse and modern technical solutions, both by officials and the local community and individuals (e.g. the use of mobile levees or taking into account extreme phenomena related to climate change in the policy). Quoting economist Peter Drucker, 'If you can't measure it, you can't manage it', in order to manage the risk of disasters, you need to estimate the elements of risk, including social vulnerability and multi-hazards.

Another aspect that emerges is the use of the term typical of tourism in this case—'tourist attraction' or 'tourist attractiveness of the place', which seems equally unethical in the context of victims (despite the fact that such a term has a theoretical basis from the scientific point of view) (Rucińska and Lechowicz 2014).

If we are evaluating someone's act, which component of the act should be taken into account and distinguished? The dominant contemporary approach in ethics is not to judge another person, but to evaluate his deed. There is no single consensus on which component of the act should be distinguished and assessed: the intention or the substance of the act itself, or its effects? This is a key question in the context of natural disasters. While the 'utilitarian' ethics and the ethics of virtue usually accept the weight of the effects, the ethics of laws and principles accept the importance of the intention or the matter of the act 'in itself'. Examples relevant to the considerations include: (i) an official deciding to evacuate people from an extreme natural phenomenon minimizes the number of victims (similarly, an official who does not make a decision to evacuate and generates thousands of victims); earthquake events in China (two different municipalities in the Yellow Bay, and two earthquakes in 1975 and 1976 including opposite impact); (ii) an official choosing lower construction costs for a building or dyke, thereby increasing the risk of casualties, the amount of material damage and long-term effects (similarly, an investment decision-maker choosing higher construction costs, thus reducing the risk of multiple casualties and high material losses and long-term indirect effects); incidents in New Orleans and weaker flood banks, or the construction of lower walls to protect against a medium tsunami in Minamisanrik, Japan, (iii) a tourist coming to Tornad Avenue in the U.S.A. risking his health or life (or a tourist coming to the site of a natural disaster risking worsening trauma in the community local). In each of these cases, we can see different effects, with different moral dimensions.

Searching for an answer to the question what kind of behaviour is appropriate for us in the context of natural hazards and disasters in the relationship between an official (decision-maker) and a citizen (local community), reference was made to ethical principles in business. In the 1990s, social and political movements began to develop, the aim of which was to restore ethical principles in the entrepreneur-consumer relationship, promoting the principle of responsibility as a key category of business ethics. By applying critical rationalism in the way of thinking, acting and moral choices and striving for social goals, which are the prevention of pain, suffering and human harm, several stages are visible: searching for the optimal solution to the problem, forecasting the effects in real life, application of theory to reality, empirical verification of results with the expected results (effects); critical analysis, assessment and discussion of the causes of discrepancies (expectations-effects); construction of a new, improved operating model (Breczko 2017).

In further search for an answer to this question, based on the method of 'social critique' (Michael Walzer) and analogies, important points were used in the stages of interpretation of the situation (Sroka 2017), which are:

(1) Determination of the 'starting point', i.e., a reference to the situation at the moment of starting the interpretation process; sometimes, these are media reports indicating the unethical conduct of enterprises it may include deceiving customers in terms of information about products—i.e., automotive companies talking about the process of CO_2 emission to the atmosphere, cold cuts—hiding additional ingredients in the product, but also financial pyramids; the effects of such actions cause suffering and sometimes death of people. Similar situations can be found in the case of natural disasters, when weaker building materials were used to construct a dam or wall;

(2) Out-of-bounds knowledge of others' social status, private relationships, commitments, and ambitions; this point seems to be as important in business as it is in the performance of governmental administrative functions, such as those related to natural hazards and natural disasters (mainly in the preparation and reconstruction stages after the incident);

(3) The subject of business ethics is interdisciplinary and requires looking at various economic and financial indicators that allow assessing to what extent unethical behaviour on the basis of economic activity affects the financial condition (losses) of companies accepting such practices. It often turns out that they do not affect the phenomenon due to the lack of consumer reaction (due to the lack of knowledge, for example, about the death of thousands of people due to some decision—as part of the entrepreneur's choice);

(4) Recognition of minimum orders; i.e., at the stage of interpretation, it is suggested to identify, define, and recognize the minimum moral imperatives. Ethics of social life is part of social ethics. If we accept that the goal of social ethics is to resist human suffering, it seems legitimate to recognize the principle of resisting suffering. Contemporary business ethics more often refer to accountability, fairness, and trust; however, the principle of not causing suffering was the first and should be continued in business and ethics, also in the area of risk management and Disaster Risk Reduction;
(5) Recognizing certain common values; it makes it possible to search for whether, and if so, how—suffering, as well as the struggle with its causes—belong to the values that are commonly recognized. Despite the fact that the crisis of trust in the entrepreneurial profession persists (but also a general crisis of trust in society, which also affects, for example, teachers and scientists—which will manifest itself in the depreciation of the value of the scientific work of a scientist or student and a depreciation of the value of master's theses, scientific articles, etc.) which in the case of business is the effect of the free market, modern society requires the citizen to 'trust' not only in people, but also in abstract systems in order to maintain prosperity (the so-called indispensability of trust). This is because the wealth of nations is created on economic and social capital. In this context, social capital is defined as the ability to cooperate for the benefit of common goals, and their realization requires trust (regardless of social position). Similarly, the question of trust arises in the context of the decisions taken by officials and the information provided to the public (i.e., natural hazard warnings) and the appropriate and timely public response to these warnings. In addition, in the part concerning the issue of consumers (ISO documents and standards), the word 'suffering' appears and the issue of human health protection and safety.
(6) A critical reflection on what is known and where we are; there is a reference to the knowledge about the economic profitability of investments (mainly cattle) and abandonment in case of extreme events, i.e., floods, hurricanes, tsunamis;
(7) Understanding how to minimize bans in specific situations; in the context of the liability of the trader causing suffering and the consumer's shared responsibility for it;
(8) Obtaining the approval of recipients of moral judgments; in the process of interpreting moral judgments, the approval of the recipients of these judgments should be obtained; how to do it? Individual recipients should be appealed to and strengthened by creating appropriate institutions;
(9) Determining the response, which may take the form of a general practice, custom or behaviour; this is difficult to establish, but it is suggested to focus on the strength and ethical sensitivity of the individual consumer. The consumer should develop the habit of analysing the purchased goods, taking into account the standards of manufacturing work and including their own assessment in the decision to purchase the goods. In this context, the analogy with DRR is very difficult because many individual actions depend on decision-makers and local government organizations before, during and after the disaster.

Each of these stages of interpretation (resulting from the Michael Walzer method) can be considered in the context of the ethics of non-suffering, as a way of assessing business practices, as well as in the context of natural hazards and Disaster Risk Reduction.

Ethical issues may be accompanied by any economic activity. Elements of business ethics are linked to actions in DRR. There is a question of maximizing profits in the company (by Milton Friedman: The business of business is business). Useful for DRR should be: (i) the definition of utilitarianism: the doctrine that actions are right if they are useful or for the benefit of a majority; (ii) the theory of consequentialism: a group of ethical standards by which the principles and actions should be assessed on the basis of their consequences (Dziadkiewicz 2012).

Anthropocentric worldviews inhibit a leader's ability to reduce suffering (Crosweller 2022). It is related with the philosophical question of what the value we place on life? (Friedman 2020). But also compassion, care, and justice are critical ethics in disaster management leadership (Crosweller 2022).

5.6 Identification of Ethics Problems in Natural Hazards and Disaster

Recently, natural disasters are the result of the dynamics of nature itself and the increase in the number of people operating in risk areas. The last twenty years' floods and storms were the most prevalent events (the number of them more than double, from 1.389 to 3.254, while the incidence of storms grew from 1.457 to 2.034). There were recorded major increases in other categories including drought, wildfires and extreme temperature events as well as a rise in geo-physical events including earthquakes and tsunamis which have killed more people than any of the other natural hazards (*The Human Cost of Disasters 2000–2019*; www.reliefweb.int). Level of losses (human and material) are linked with social vulnerability to natural hazards.

It is worth noting that ethical issues pertain to human wellbeing and we have a moral obligation not to harm others and to help those in distress. The Globe is shrinking, and the distance between continents in the West of Asia is more than geographical in a global age of communication (Zack 2009).

Moreover, ethical problems have different spatial and social scales. They may concern a group of countries, a country, a region, a city, or a village; likewise, an office, institution, organization, community, family, or individual. Behaviours and actions are related to regional or local culture and historical experiences, the influence of the political system, as well as the existence or lack of systemic solutions and procedures of state or private organization of employees of a given institution, usually in a decision-making position. The ethical problems associated with natural hazards and natural disasters are incompetence and ignorance (Alexander 2017), as well as insensitivity or indifference (Rucińska 2017), selfishness, self-focus, and the corruption.

Geoethics cover the issues of natural hazards and disasters which, in turn, result in ethical problems of various scales. Identifying these is important when investigating specific problems, such as any ethical or unethical behaviour in the case of natural hazards and disasters. This is about behaviour and actions at various social levels that are played out by the individual or within the local community; there is the cultural role being performed, along with historical experiences. And then there is the institutional level where a systemic problem may occur as a result of state organization or an employee of a given institution, usually those in a decision-making position.

5.6.1 Corruption and Natural Disasters

The phenomena of corruption and the black market, incompetence and ignorance, as well as the exploitation of social groups that accompany natural disasters around the world (Alexander 2017), are a major problem. These most often appear in the case of earthquakes (Escaleras et al. 2007; Ambraseys and Bilham 2011) and also tsunamis (Kushida 2012) although, at the time of writing, the geoethical effects of the impact of the 28 October 2018 Tsunami in Indonesia have not yet been assessed. Criminal groups such as the mafia can expedite their involvement at the same time, along with propagation of the drug trade (Bales 2007). One example of this could be the Bam earthquake in Iran, in 2003, and the subsequent widespread problem with heroin addiction (Tai 2006).

Corruption has a negative impact on sustainable development and sustainable mitigation of natural disasters (Ziervogel et al. 2017). One example concerns the interaction of criminal organizations, political forces, and measures that were introduced to fight floods and landslides in San Salvador. There, physical measures and assistance in disaster reduction did not go to those areas that needed it most, but to those that were marginalized and depraved (Wamsler et al. 2012). There is a lack of correctly coordinated use of humanitarian aid such as in earthquakes, hurricanes, or droughts, where the arrival of such aid ends up being an opportunity for corruption to flourish (Breau and Carr 2009).

The issuing of funds taken from humanitarian assistance should be accompanied by some form of spending on the need for protection, so that funds from humanitarian aid can be spent as intended, i.e., to protect areas against flourishing corruption (Walker 2005). Irregularities were reported repeatedly by the press in the case of the New Orleans flooding as a result of Hurricane Katrina in 2005. The city mayor, Ray Nagin, was convicted by the federal commission for accepting bribes in order to rebuild the city after Hurricane Katrina. There is a lot of reported press as regards global scandals related to natural disasters (which can be done so on sites such as talkingpointsmemo.com, IPF 2015). What is noteworthy, however, are these words mentioned by New Jersey Governor Chris Christie during a Fox News broadcast:

'What matters is ... people waited over 65 days for federal relief aid ... during Sandy. That was six times the amount of time they waited after Hurricane Andrew and ten times more than they waited [after] Hurricane Katrina' (www.fox.news 2017). Whether this statement complies with the facts is something this author has been unable to verify, but it is important in the context of geoethics.

Certain criminal penalties for misdemeanours or fraud in the U.S.A. are systematized work as regards prosecutable offences related to the catastrophe. The Disaster Fraud Task Force (2011) (formerly known as the Hurricane Katrina Fraud Task Force) for the Fiscal Year of 2011 relates to Hurricanes Katrina, Rita, and Wilma in 2005 and is continuing to build up a file of prosecution cases related to these disasters. It operates in partnership with the American Red Cross, and a good many private sector organizations that have been assisting law enforcement agencies in identifying new hurricane-related fraud schemes. Forty-seven U.S. Attorney Offices across the country charged 1439 people in 1350 cases, with various fraud-related crimes that had stemmed only from Hurricanes Katrina and Rita, and many of these defendants received criminal sentences. One of the indispensable components of that success has been the National Center for Disaster Fraud (NCDF), formerly the Hurricane Katrina Fraud Task Force Joint Command Center (Report 2011, DFTF).

The subject of geoethics is discussed at numerous conferences around the world: among others, by the IAPF, 'Ethics and public life' pt. Indifference and sensitivity and public life in Krakow in 2015, and 'Disaster Risk Reduction' Conference in Warsaw in 2017.

One of the problems are the methods the media use when reporting on threatening situations, especially on television. Sometimes they do serve to provide the public with information or with education, but others often use the station as a tool to primarily promote their organization and make themselves more attractive. Similarly, the use of a crisis situation in the Public Relations process has been observed in the circles of political and governmental dignitaries, who then choose to make promises that are often of an electoral nature.

While corruption is regarded as an 'old phenomenon', by comparison, dark tourism is comparatively recent and can be described as being the commercialization of locations that have been threatened and affected by disasters. As such, these include offering trips to locations and places at risk (otherwise known as dark tourism) (Rucińska and Lechowicz 2014), making use of documented images of the disaster to promote sensationalism, increased viewership, company or individual advertising, or even as a form of artistic promotion (Rucińska 2016).

Ethical issues are saddled with individuals that are forced to individual decisions in the face of danger or as a result of their roles. Making a choice depends on the level of acceptability of risk, which is either an individual feature or the result of business procedures or thereof. As topic, acceptability of risk in the context of natural hazards has not yet been thoroughly examined and elaborated. It can be said, however, that acceptability is related to the characteristics of a given culture, as well as to individual characteristics, including empathy. In the following chapters it will be looked at the universality of the concept of reducing the risks of natural disasters.

One of the wider problems is corruption. The possibility of corruption developing during a natural disaster (Alexander 2017) as well as the COVID-19 pandemic (Stefanile 2020) is emphasized. Corruption is a major factor in loss of life from earthquakes (Ambraseys and Bilham 2011).

Corruption is discussed in many fields. There are a many different definitions of corruption. These definitions vary according to cultural, legal, or other factors. There is no consensus about what specific acts should be included or excluded. In practice, definitions of corruption are often too general (*OECD Glossaries. Corruption* 2008). The classic definition, in the 1990s, is the misuse of public power for private gain. 'Corruption involves behaviour on the part of officials in the public sector, whether politicians or civil servants, in which they improperly and unlawfully enrich themselves, or those close to them, by the misuse of the public power entrusted to them'. Transparency International developed the National Integrity System approach as a comprehensive means of assessing a country's anti-corruption efficacy sector by sector in countries. The National Integrity System evaluates key 'pillars' in a country's governance system, both in terms of their internal corruption risks and their contribution to fighting corruption in society at large. It was assumed that if some or all of the pillars wobble, these weaknesses could allow corruption to thrive and damage a society. The pillars analysed in a National Integrity System assessment typically include: legislative branch of government, executive branch of government, judiciary, public sector, law enforcement, electoral management body, ombudsman, audit institution, anti-corruption agencies, political parties, media, civil society, and business. A National Integrity System assessment examines both the formal framework of each pillar and the actual institutional practice. The analysis highlights discrepancies between the formal provisions and reality on the ground, making it clear where there is room for improvement. The analysis is undertaken via a consultative approach, involving the key anti-corruption agents in government, civil society, the business community, and other sectors (www.transparency.org).

The most important issues at a conceptual level, corruption is a 'multi-layered phenomenon' (i.e., in management of corporation are non-ethical decision or action in planning, organization, motivation, and control). Usually, it is depicted as behaviour based on the individual's motivations for engaging in corrupt behaviour (Rose-Ackerman and Søreide 2011; Breit et al. 2015). It should not only be regarded as a state of misuse, but also as 'a process of gradual institutionalization of misbehaviour which contributes to legitimizing behaviour and socializing others into it in such a way' (Ashforth and Anand 2003).

There is evidence that inefficiency is one of the causes of corruption (Anderson and Gray 2007; Ferreira et al. 2007). The combat with corruption relates to difficulties in measuring it empirically due to its clandestine nature. Many corruption measurement methodologies use perception, which is of questionable validity and reliability (WCO Research Paper 2010). 'Corruption is not only an economic phenomenon, but also a moral one'. It should be noted that many countries affected by a disaster or conflict are classified in the higher corruption group (this refers to Transparency International's own Corruption Perceptions Index (CPI)) (Walker 2005). Defining corruption too broadly and the lack of a universal measurement is a problem. Moreover, the use of the Corruption Perception Index (CPI) as an instrument to assess the effectiveness of anti-corruption measures in the country is a mistake, because this measure does not allow determining the reasons for changes in its size (Lewicka-Strzałecka 2018). Public corruption and measuring are discussed (Yamamura 2013, 2014).

The phenomena of corruption and the black market, incompetence and ignorance, as well as the exploitation of social groups accompanying natural disasters around the world (Alexander 2017) are a big problem. They most often appear in the case of earthquakes (Escaleras et al. 2007; Ambraseys and Bilham 2011), also tsunami (Kushida 2012) (Table 1). Criminal groups as mafias can expedite their involvement. In the same time, and the propagation of the drug trade (Bales 2007). An example of this could be the Bab earthquake in 2003 and the wide problem with heroin addiction (Tait 2006).

Finally, if you can't measure it, you can't to improve it, by philosopher Lord William T. Kelvin.

5.6.2 Corruption and Poverty

The corruption deprives the poorest, by which poverty is exacerbated and, at higher social levels, it may keep undeserving elites in power (Hoogvelt 1976). As Johnston (2009), online, after Lewis has said 'The links between corruption and poverty affect both individuals and businesses, and they run in both directions: poverty invites corruption, while corruption deepens poverty'. Foreign aid includes DRR and other vulnerability reduction measures. If the 'dirty money' was invested in the communities where the money came from, perhaps there would be no need for foreign aid to initiate and support DRR (Lewis 2012). Poverty generates further pathologies, such as slavery or prostitution (Bales 2007), drug addiction (Tait 2006), and trafficking in drugs (Bales 2007), and consequently creates benefits for criminal groups and their enrichment (Wamsler et al. 2012). This leads to increased activity and contacts with the state administration (Alexander 2017).

Institutional racism is a pervasive, but often subtle, social problem, a condition, i.e., as segregated housing, as exclusion and discriminatory harms in their dealings with institutions or individuals. Poverty and lack of education may be the result of intergenerational disadvantage that persists in groups of African Americans and other non-whites in the U.S.A. (Zack 2009). The problem is that from the very beginning, weaker social groups are located in inferior or endangered areas (houses are built there for them).

Moreover, the non-profit organizations (NPOs) are instrumental in blurring the boundaries between humanitarian and political responsibilities in neoliberal regimes. Authors focuses on the influence of private actors, such as non-profit organizations (NPOs) and firms, which has been increasing in disaster governance and politicizes the entangled relations between NPOs, states, and disaster-affected people. There is diversity in liberal states which have two types of responsibilities in disasters: humanitarian and political using the Rawlsian approach. Disaster governance arrangements under neoliberal regimes are structured around the division of humanitarian and political responsibilities that can be lost where the different states, private actors, and disaster-affected people operate without the coordination or separation of their roles and tasks:

(1) Political responsibility when basic political responsibilities of the state are privatized and delegated to NPOs and firms.
(2) The humanitarian responsibility as a disaster offers an opportunity to nationalize and extend the political control of the state into the traditional areas of private life.
(3) The location of the basic institutional boundaries between the public and private spheres of society may be blurred.
(4) Empty spaces of responsibility between the political and humanitarian responsibility may be produced or filled in a way that some people can be increasingly excluded or included in a society.
(5) There may be a situation of overlapping of humanitarian and political responsibility where the different institutions, organizations, and individual actors operate without coordination or separation of their roles and tasks. Before, during, and after the hurricane of 2005 in the U.S.A, the marginalized citizens affected by the hurricane were losing their political citizenship and becoming increasingly dependent on the humanitarian support of the NPOs, religious groups, and the like. This shows that disaster governance arrangements relying heavily on private actors may lead to the situation of overlapping humanitarian and political responsibility. In the situation, the humanitarian responsibilities can be delegated to specialized private organizations such as NPOs or firms; political responsibilities should remain real and in the hands of states (Meriläinen et al. 2020).

An example, in the disaster governance arrangements that unfolded after the 2010 earthquake in Haiti, the American Red Cross (ARC) contributed to the mixing and blurring of the boundaries between political and humanitarian responsibilities. The humanitarian responsibility is politicized and political responsibility privatized has various adverse effects, as they: (1) leave marginalized people vulnerable, (2) transform NPOs into political agents of liberal governments and undermine the trust in their political neutrality, and (3) make the political structures of the host countries of the disasters more burdened in the process (Meriläinen et al. 2020).

5.6.3 Corruption and Construction

The most examples of immoral activities exist in relation to construction, and this in turn is most often associated with regions of seismic activity—earthquakes, as well as with tsunami (Escaleras et al. 2007; Ambraseys and Bilham 2011; Kushida

2012). Why? There are several reasons. Construction takes place within strict time frames, and exceeding them results in huge financial penalties; therefore, regular checks and confirmation of correctness should be performed at individual stages of the construction, from design, material specification, tender to implementation (Lewis 2003). The construction time regime does not, however, explain the unethical behaviour of engineers and contractors turning a blind eye to inaccuracies and errors and then taking these buildings and confirming that the required conditions and rules are met. Design standards can be the same as or higher than the standards required by legislation—but not lower. Building law should meet a level that cannot be exceeded (Lewis 2003).

There are many examples that legislation does exist, but supervision and control have been inadequate or absent at all. An example is the 1999 earthquake in the Marmara region of Turkey, where there was widespread destruction of buildings despite the provision of funds for the construction of earthquake-resistant buildings. Irregularities in the mechanisms for controlling the work of local construction contractors in communes meant that many buildings were not built according to construction standards (Ozerdem 1999).

If this is the case in developed countries, how is legislation expected to work effectively in developing countries? The improved construction of a building for housing of all kinds should also be part of special programmes; improved building design should be achieved in all housing activities and other sectors; require: the creation of long-term programmes for the improvement of legislation, training programmes and integrated demonstration projects, special information for non-professional use and its dissemination. The need of this type is so universal and pervasive it requires cyclical repetition of projects and their continuous continuation.

It happens that political and economic decisions contribute to a natural disaster through the links between responsibilities, opportunities and pressures, which are consequently a violation of human rights. Examples include the three earthquakes in Turkey in 1999, 2003, corruption, and state neglect. An example of unethical actions are: organizational deviations ('systemic corruption; state collusion with corporate crime; government collusion in the illegal activities of its own elites; war crimes; negligence; and post-disaster cover-ups'); economic policy (according to the author, liberalization economics in 1989 and globalization) (Green and Ward 2004), and the corruption of political spheres (Green and Ward 2004), which developed during this period: repeal of restrictive building rules (Green and Ward 2004), illegal building, i.e., by adding a storey, and constructing amnesty for construction. The effects of these practices, as well as ignorance for professional knowledge (engineering and geological), deliberate lack of interest in the knowledge of ruling zones and planners, the culture of laissez-faire and populist clientelism, injustice of the system in the casualties and losses, including the earthquakes in Marmara and Duzka in 1999 and Bingöl in 2003. Poverty, corruption, and the authoritarian nature of the state are the main features in predicting and assessing vulnerability to this type of natural disaster, as well as the effects of war and politically generated urban migrations combined with poverty associated with the Bingöl tragedy. There are some similarities in terms of politics, corruption, organized crime, reconstruction, and the lack of government accountability from earthquakes between the regions of Italy and Turkey (Green and Ward 2004). 2010 Elazığ earthquake (Elazığ earthquake in eastern Turkey, M 6.7; at least 41 people killed and more than 1.600 were injured. The shocks were felt in neighbouring provinces. In Turkey and in neighbouring countries like Armenia, Syria, Iran, Georgia, and Lebanon, many aftershocks have been reported) argues that despite significant advances in seismic safety elsewhere in the country, low-economic rural areas continue to pose a risk of life, particularly schools. Buildings that have been damaged most require building inspection (Akkar et al. 2011).

In the U.K. and the U.S. construction industries, wealthy nations have a long history of construction legislation. Often, changes take place as a result of the latest catastrophe, and this process is continuous (i.e., *Natural hazards Observer* 2002). Despite this, failures and corruption continue to occur, including in developed countries. This indicates the universality of the problem and the lack of solutions in the field of geoethics. It should be remembered that the need for fast services in reconstruction reduces the quality control of services and finally quality of new constructions (Sorensen 2014).

The analysis of corporations in the EU (EU28) countries shows the highest corruption in the following sectors: (i) construction and building, (ii) health care and pharmaceutical, (iii) engineering, electronics, and mot. vehicles (*Flash Eurobarometer* 457). This can be confirmed by the incidents of earthquakes in Turkey and the pandemic. In 2020, during COVID-19, there were cases of vaccination outside the established rules of vaccination sequence in countries, i.e., in Italy (100.000 people were vaccinated and 200 people were vaccinated in Poland), which was widely reported in the media in December 2020. During the pandemic, maybe there will also be illegal sales of vaccines on the black market, as well as sales that do not comply with previously concluded contracts.

Lack of appropriate action by officials—as a consequence: different parameters of the height and width of the adjacent flood embankments on the Vistula in two adjacent administrative units in Poland (on the border of Podkarpackie and Świętorzyskie Voivodeships, Tarnobrzeg-Sandomierz); cheaper levees (New Orleans), tsunami protective walls (2011)—and structures resistant to extreme phenomena, greater material damage and more victims.

There is a strong need to objectively assess corruption. It resulted in the creation of a measure of the difference between the measure of the size of the physical public infrastructure and the cumulative price that the government pays for public capital. It has been found that if the difference (between the money spent and existing physical infrastructure) is greater, then corruption is greater (more money is spent on fraud, bribes and embezzlement). The measure was established for the administrative units of Italy (20 regions) in the mid-1990s, controlling at the regional level possible differences in the costs of public buildings. Then, when comparing construction costs in the public and private sectors, it was found that their spatial distribution is varied and that in the south of the country, in terms of infrastructure, low costs for the private sector are accompanied by higher public sector expenditure. It is in the south of the country that public sector contracting and the ineffectiveness of the construction industry are particularly common. The created model (and a map) is potentially useful for studying the differences in the causes and consequences of corruption in the regions of the country (Golden and Picci 2005). The timeliness of the need to evaluate the intensity and dynamics of corruption is also emphasized by other corruption evaluation studies based on the indicator of the decline in socio-economic development in Guatemala (Estrada et al. 2018).

5.6.4 Corruption and Humanitarian Relief

There is a lack of the intended use of humanitarian aid, for example in phenomena such as earthquakes, hurricanes, or droughts, where the arrival of humanitarian aid is an opportunity for corruption to flourish (Breau and Carr 2009). The catastrophe and the crisis (after the catastrophe) generated by the tsunami are conducive to corruption (Kushida 2012) and humanitarian aid abuses. Moreover, the subsidies themselves can increase social inequalities and, like political inequalities in power, lead to corruption. An additional burden is the fact that the spending of humanitarian funds is accompanied by a strong need to spend other funds in parallel to protect funds obtained from aid to ensure that they are spent as intended, i.e., to protect against burgeoning corruption (Walker 2005).

During a disaster, when money arrives from various sources and via various routes (often millions of dollars in revenues within a few days), it is necessary to create management and audit systems to protect against corruption. Corruption arises for two main reasons:—exclusions when individual recipients of gifts have already left their place of residence;—inclusion (inclusion)—when people with help are directed to a place where in fact help is not needed (i.e., it is no longer needed, because the population was evacuated or fled). All activities then require additional time and the relocation of the fund-raiser. Consequently, delayed deliveries with assistance create a surplus of (products) to local needs. In addition, aid agencies build their own operating systems (supply, service, etc.) that are parallel to the existing local systems, while strengthening them; however, they also result in the accumulation of people and employment of (selected) educated people who know languages and computer services, who are offered remuneration rates much higher than local rates for the time related to the assistance, which is good in the long term. In addition, the influx of people increases food prices and house rental prices. Agencies give a positive result of implementing a new system in a given local community, but these activities usually do not permanently affect the solution of the problem on the local labour market. This is often due to the need to quickly employ a large number of local workers—culturally and religiously different, with complex ethnic relationships (which leads to tensions and conflicts for which aid organizers are not prepared). The aid operation of an inexperienced agency is highly vulnerable to corruption (Walker 2005).

Alongside 'dark tourism' and humanitarian aid, volunteering is developing.

Volunteer tourism is commercialized market that meets the demand for a different travel experience for the more morally conscious traveller, while at the same time it provides opportunities for economic gain for the organizations that act as brokers of such experiences. This interaction raises several ethical issues in terms of serving a mission while making economic gains. In general, there is an acceptable relationship between monetary gain and altruistic service, within the context of enlightened self-interest provided that the beneficiary of economic gains diverts profits into serving their mission (Tomazos and Cooper 2011). It is observed that a rise in international medical volunteering (IMV) poses complex issues for organizations, clinicians, and trainees to navigate as well as ethical implications of IMV, such as scope of practice, continuity of care (Sullivan 2019).

5.6.5 Corruption Versus Sustainable Development

Corruption is a major factor in loss of life from earthquakes (Ambraseys, Bilham 2011), and there are important lessons for risk managers and the international development community (Hill 2012). Corruption has also a negative impact on sustainable development and the lasting mitigation of the effects of natural disasters (Ziervogel et al. 2017). One example concerns the collaboration of criminal organizations, political forces, and measures to combat floods and landslides in San Salvador, the capital of El Salvador, where physical measures to reduce disasters do not go to the places most needed but are marginalized and depraved (Wamsler et al. 2012). There is no proper (intended) use of humanitarian aid, i.e., in phenomena such as earthquakes, hurricanes, or droughts, where the arrival of humanitarian aid is an opportunity for corruption to flourish (Breau and Carr 2009). The phenomenon of spending humanitarian aid funds is accompanied by the necessity to spend other resources on protection, so that humanitarian funds are spent as intended, (i.e., on protection against the burgeoning corruption) (Walker 2005).

The irregularities were repeatedly reported in the press, including in the floods and Hurricane Katrina in 2005 (Table 6.1). In New Orleans, a federal commission of accepting bribes to rebuild the city after the hurricane convicted the mayor of the city. In the context of this event, the words: '*What matters is ... people waited over 65 days for federal relief aid... during Sandy. That was six times the amount of time they waited after Hurricane Andrew and ten times more than they waited [after] Hurricane. Katrina*' (Fox.news 2017). There are a lot of press releases reporting on global disaster-related scandals (i.e., talkingpointsmemo.com, IPF 2015). Such reports are important in the context of geoethics. A systematic work on the prosecution of crimes related to disasters is a certain compensation for offences or abuses in the U.S.A. For example, the Disaster Fraud Task Force (formerly known as the Hurricane Katrina Fraud Task Force) referred in its 2011 Fiscal Report to Hurricanes Katrina, Rita and Wilma in 2005. About 1350 different fraud cases were identified. The sources of which were two hurricanes—Katrina and Rita. Many of the accused received sentences. The group continues to prosecute disaster-related cases in partnership with the American Red Cross and many private organizations that assist law enforcement in identifying new hurricane-related fraud programmes. The success was due, among others, to National Disaster Fraud Center, NCDF, formerly the Hurricane Katrina Fraud Task Force Joint Command Center (Report 2011, DFTF).

Two situations should be distinguished: building in an area where there should be no building, and building without following the rules (here we can see an analogy in the case of floodplains in Central Europe, including Poland). Geological, topographic, and geographical analyses are used to determine the areas of earthquake risk and to create micro-zones for the purpose of spatial development planning. In Peru, for example, they are used with great success and are based on observations of natural surface features (Kuroiwa 1982, 1986, 2002). 'It is better to invest in normality than in a catastrophe' taking into account adjustments in the development strategy:

- All economic sectors must take into account the earthquake, not just those for civil defence and rescue (such as sensitive infrastructure), especially for construction, including housing.
- In addition, high-density populations need a different balance of risk as they need to be dispersed over time. Therefore, not only resettlement should be required, but also redistribution of, i.e., hospitals, schools, and services—social services of all kinds. This approach will be socially fair and commensurate with current and future urban and rural populations. Although we hear about national disasters, we should always remember that they are made up of local disasters and that we need to act on the municipal level (Lewis 2003) (Table 5.1).

Table 5.1 Corruption and irregularities' examples

Marmara, 1999	The U.S.A., 2005	Japan, 2011
There is legislation but there is no control, or control is inadequate (Ozerdem 1999, Lewis 2003).	Public employees were accused of bribes from relief-funded contractors and of overbilling the government. Hurricane Katrina in 2005 (Leeson and Sobel 2008).	The misuse of reconstruction funds was revealed in the case of the Great East Japan Earthquake. It was reported 'a special account budget to fund the reconstruction of communities devastated by earthquake, tsunami, and nuclear disasters has been used to pay for unrelated projects'. Some money earmarked for reconstruction work was spent improperly on projects to improve the earthquake resistance in buildings of the central government (Yamamura 2013).

Source Author's own elaboration

5.7 Selected Ethics Issues—A Review—the Challenges for Disaster Risk Reduction

There are many examples of corruption, including those described by scientists: (i) San El Salvador: One example concerns the interaction of criminal organizations, political power, and measures to combat floods and landslides in San Salvador, the capital of El Salvador, where physical measures to reduce disaster they do not go to places of greatest need; (ii) China's corruption costs have been estimated as 8% of spending from 1996 to 2005. The destruction of so many schools in the 2008 Sichuan earthquake, which killed 75.000 people including 900 children in one school, was strongly suspected to have resulted from corruption in construction (Lewis 2008); (iii) The Chinese government states that it is battling corruption; nonetheless, 'corruption remains rife and is one of the most potent sources of public anger' (Branigan 2011). Weakness points of DRR need analysis because of non-ethical decisions and actions or inactions that affect the dimension of natural disasters (Tables 5.2, 5.3, and 5.4; Fig. 5.5).

5.8 Promoting, Discontinued, and Corruption

It should be noted that there is a strong link between large-scale charity activities and the activity of the mass media (mentioned above, concerts, exhibitions, activities of government, or party representatives). This type of activity includes an element of conscious advertising (actors, artists, politicians appearing in the performance). It is difficult to assess the line between helping and being active, resulting from social sensitivity, and preserving its appearances and engaging in self-advertising, which would mean the opposite phenomenon—social indifference. Examples include hundreds or millions of products (T-shirts, gadgets) with the names of bands, artists, parties (inscriptions on the means of transport, etc.), as well as television broadcasts, interviews, in the media supported by films or photos. In part, it can be both. It should be emphasized, however, that the concert hosting around 500.000 USD for Haiti during one cultural event is without doubt a significant financial aid in the post-disaster period. Likewise, it is difficult to judge the arrival of politicians in the disaster area, giving a speech and taking a photo of the devastation. A similar question arises whether they were performed in order to help, document work duty or professional activity. Usually, this is accompanied by the promise of help and reconstruction. This activity can also serve to improve the political image of voters (personal or institutional). This question is justified when media appearances are not followed by an unequivocal solution to problems, including: refining the principles of public information and warnings, preparing for subsequent events, developing target and long-term solutions in spatial planning, taking into account natural hazards and the variability of nature, i.e., possible more frequent extreme hydro-meteorological phenomena, development and implementation of strategies for adaptation to threats,

Table 5.2 Important issues affecting the amount of losses, requiring recognition and solutions

Building and evacuation	Moral hazard	Insurances	Degradation of the environment	Self-promotion and self-realization	Corruption
Building poor houses on risk area. Not using International Building Code (IBC) and others. The need for quick services reduces the quality control of these services. Build protection for medium-risk, not extreme-risk events. Reconstruction of buildings, services, industry in the same risks areas (no relocation). There are no escape routes. Not solving traffic jams on escape routes. Pedestrian evacuation routes to safe points that do not take into account ethical aspects related to the necessary distance to be covered by the weakest people at a given time (i.e., tsunami).	It can happen, when a person takes more risks because someone else bears the cost of those risks. Moral hazard is a situation in which one party gets involved in a risky event knowing that it is protected against the risk and the other party will incur the cost.	No financial guarantee from the state for insurance companies (reinsurance). Fixing the border of a dangerous phenomenon (i.e., strong wind) in such a way, in order not to cover the higher number of similar events. A situation where it is impossible to prove a storm or tornado by the injured person due to the lack of an appropriate monitoring of hazards network at the local level. Insurance conditions presented in an incomprehensible way. Excessive insurance premiums for individuals. Insurance conditions do not guarantee the return of investments, i.e., in agriculture for farmers.	Lack of protection of mangroves and coral reefs in the tsunami risk area. Practices devastating the natural environment on the coast, which increase the risk of natural disasters due to the deterioration of the quality of the natural environment.	Mass media. Authorities (Politicians). Tourist companies. Dark tourists. Individual volunteers.	Verification of authorities' decisions: Only short-term decisions; developers' action and politicians cooperation; no control of services quality. Action based on economic profits: The choice of cheaper solution and no community profits with quality of security. Misuse of money or products collected by charity organizations.

Source Author's own elaboration

Table 5.3 Examples of gaps

Gaps
Legislative obstruction as well as bureaucracy limiting action for DRR. Lack of good relationships and knowledge exchange: universities—local governments—authorities—compliance. Low interest in new knowledge among public administration employees. Lack of informal education, i.e., about the local evacuation system. No local educational and adaptation activities to natural hazards. Information on threats in the commune is insufficiently public. There are no rules for relocation from risk areas. Lack of rules for the obligation to implement modern technological solutions after the catastrophe (abandoning the usual reconstruction and implementing modernization of buildings, etc.). There are no rules for the public to verify the actions taken after the disaster by the authorities. The inefficiency of the authorities is supplemented by the spontaneous, heroic work of the inhabitants during the defeat. Lack of procedures (solutions) for strengthening a new local bond after a disaster, i.e., in order to deepen knowledge and awareness of threats. No protection for those affected by the disaster.

Source Author's own elaboration

Table 5.4 Framework of public choice theory

The framework of public choice theory
Government is anticipated to play a leading role in reconstruction and allocates a budget for that purpose. In this case, various groups related to public works attempt to receive orders from the government. Because of information asymmetry or the support of favour-based politicians, groups are able to seek benefits even though their works are not associated with reconstruction (Yamamura 2013).
The occurrence of disasters gives politicians an incentive to misallocate disaster expenditure in order to increase the probability of their re-election. Consequently, this actions stop help to those who need it most (Sobel and Leeson 2006); (Yamamura 2013).
Lack of implementation of law for rational spatial planning and development for profits because of land value. Population in urbanized areas should be dispersed. There should be resettlement and redistribution of hospitals, education services (Lewis 2003).
No verification of authorities' decisions. Only short-term decisions are taken instead of long-term actions. Developers' action and politicians' cooperation should be evaluated due to control of developers' services quality.
The authorities choose technical solutions based on economic benefits: they choose a cheaper solution and do not guarantee the highest quality of safety for society.
The embezzlement of money or products collected in shares by charity organizations.
Officials do not have the necessary knowledge to make decisions about the number of evacuation sites and the choice of tsunami vertical-evacuation (TVE) sites or escape routes. What is needed is a spatial analysis model that takes into account ethical criteria (Engstfeld et al. 2010; Wood et al. 2014).

Source Author's own elaboration

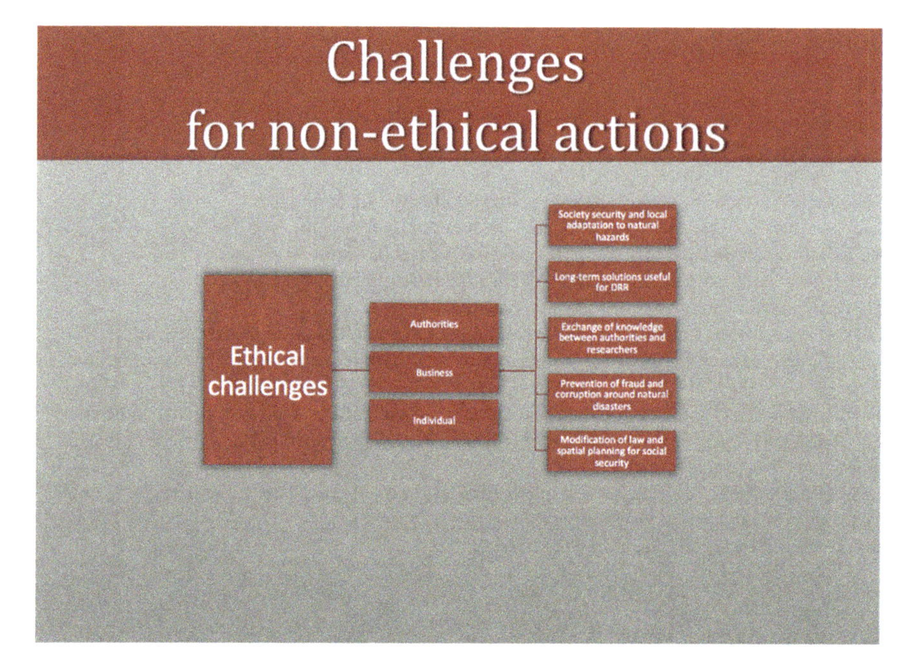

Fig. 5.5 Challenges of the twenty-first century. *Source* Author's own elaboration

solving solutions based on permanent relocation from hazardous areas to safe areas, continuous development of social awareness and preparation for making such difficult decisions, raising awareness of the need to insure, developing guarantees for insurers in the event of a natural disaster.

Research on the perception of natural hazards by schoolchildren in Poland indicates the need to develop formal and informal education about natural hazards and extreme phenomena (Rucińska 2011; 2012a, b). The element of education should also be the potential risk of losses (life, property, and social values) in various spatial scales and the principles of reducing this risk, including the ethics.

Solving problems sometimes requires legal changes, taking into account the negative impact on the administration budget, i.e., in the case of changes in the value of land as a result of its reclassification—undoubtedly the involvement and actions of local, regional, and state authorities.

While funds often reach the affected, there are often no further preventive action and preparation of the area and people for further events. Lack is of follow activities that are the solution of problem. Such a testimony is the flat of the flood victims from 1997 in houses of light construction donated by the Danes last year, which were to serve the victims immediately after the defeat and serve for over 20 years, exposing them to further health hazards due to frequent failures and fires in the electrical system, moisture, and mould. These donated houses were previously used and are in fact many years older than the flood of 1997 and the estate of St. Catherine (commune of Siechnice)(Sobala 2015, www.wroclaw.tvp.pl), which has existed for years without the problem being resolved by local authorities.

These 'humanitarian homes' have become homes for decades. This approach of the authorities is an expression of indifference towards citizens, and this accompanies:

- Apparent rationality, economy, and pragmatism (quality of selected solutions versus costs).
- Using the situation (circumstances) to build a political image.
- Willingness to maintain a job in public administration ('willingness to survive', term of office in exercising power).
- Disaster relief and occasional relief (financial aid for one-off benefits, reconstruction usually at an emergency, humanitarian aid) are a short-term solution that does not benefit long-term DRR policy (Fig. 5.6).

In the case of politicians, failure to solve the problems surrounding natural disasters is caused inter alia, by the term of office of the authorities (relatively short term of office in relation to the time needed to solve a problem resulting from a hydraulic threat or a disaster). The authorities are aware of the difficulties they will face when undertaking the challenge, at the same time without guaranteeing the expected result during the term of office, and often omit these complex, but also the most important measures to improve social security. That is a why services undertake the intervention and aid during the incident, the disaster and is limited to transferring certain funds

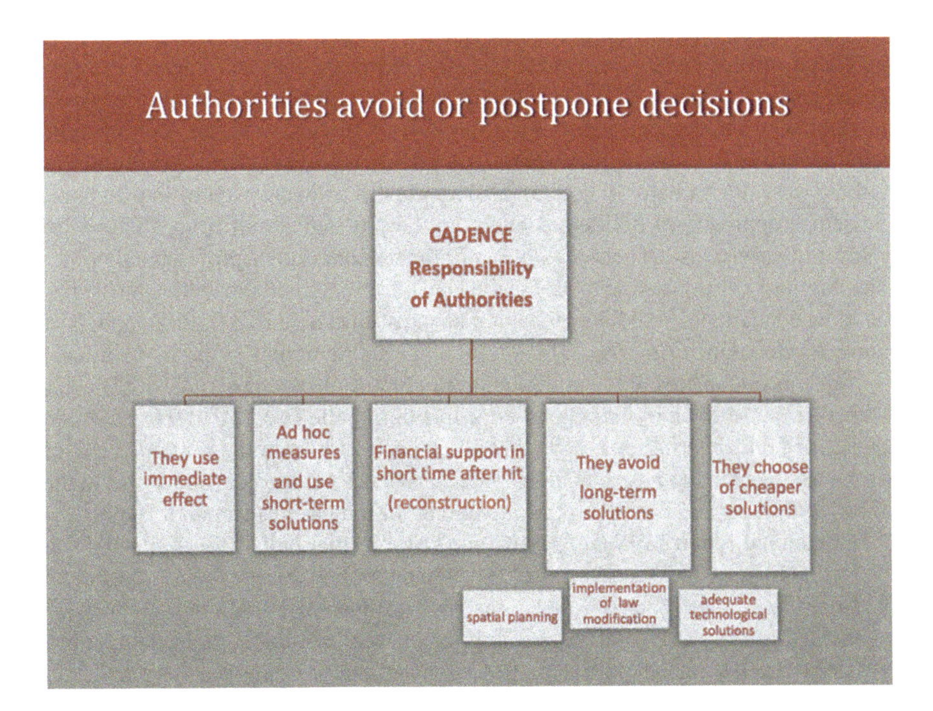

Fig. 5.6 Cadence and decisions-making. *Source* Author's own elaboration

for fast reconstruction after it. It is very easy to find on the Internet documentation of the mass media proving the presence of politicians at the crash site, indicating their interest in the case, regardless of the country, which indicates the universality of the phenomenon and the problem. However, it is much more difficult to find a presentation of solutions adopted and implemented after the disaster.

State and local governments are making efforts to help the population recover from disaster. Special national and international funds are mobilized for this purpose—the EU Solidarity Fund, which shows the sensitivity of the authorities to human suffering. However, it should be emphasized that the failure to make any efforts to regulate many essential matters for minimizing the effects of successive natural hazards indicates indifference to potential victims and only the desire to survive in a convenient social position (professional and political prestige) and financial security (a satisfactory income).

Among the factors influencing the dimension of a natural disaster, it is selfishness in public expenditure and corruption resulting from greed, the abuse of political or commercial power, mismanagement, and incompetence. These features and behaviours illustrate the indifference of political spheres and companies toward society and focus on one's own interests. Corruption can take place in different regions of the world. It may take the form of discontinued, failure to take appropriate actions in the appropriate period or the choice of decisions and solutions saving the budget or profits, at the cost of social and economic losses suffered by the local community during a natural disaster. Choosing an economically rational choice for the local budget (i.e., choosing a cheaper technological solution that protects the population against moderate natural hazards, but cannot protect against extreme phenomena in the age of climate change or in the face of extreme phenomena) over the number of deaths, depriving entire families of shelter, work and condemning them to function for many weeks (or even months or years) in extreme conditions, or until recovery after a natural disaster, also appears as an expression of indifference. The selfish approach of politicians and corporations may result from a desire for profit or to survive as long as possible in optimal economic or professional conditions. Similarly, it is possible to assess the lack or only limited preventive measures, including educational activities on natural hazards. Inspections (i.e., in Poland, *NIK o inwestycjach powodziowych* 2014) reveal many shortcomings in the work of local governments and in spatial planning. In their light, the question becomes justified as to whether the heroism of helping other people during the floods of 1997 or 2010 in Poland, and endangering one's own health, is necessary from the point of view of crisis management and risk reduction strategies. Repeated disasters and losses in the following decades indicate that preventive long-term measures were not taken.

The national report in Poland (NIK) showed also that only 12% of the areas at risk of flooding are included in spatial development plans, and only a third of them introduced investment bans or restrictions on investment in areas at risk of flooding. It is related to the duties and responsibilities of local governments in the country (*NIK* Report 2014 after the flood in 2014; www.gazetaprawna.pl).

Considering the deontology approach, the general question is whether extreme circumstances permit abandoning of absolute rules? How do we know if, when, and how we can modify it? 'What persons in authority intend to do and carry out in disasters is an ethical matter because it involves human wellbeing. Both preparation and response require plans, and both kinds of plans have ethical aspects. Is there an ethics of disaster-preparation planning distinct from an ethics of disaster-response planning?' (Zack 2009). Existing ethical principles are because they are based on values that, in the Western tradition at least, are the result of millennia of religious and humanistic study and practice. These principles and values are the heritage of a democratic society. Ethics problem are not only in developing countries and ethical principles that exist because they are based on values that, in the Western tradition at least, are the result of millennia of religious and humanistic study and practice. These principles and values are the heritage of a democratic society and should serve as ethical guidelines for disaster in the following ways:

A. 'we are obligated to care for ourselves and our dependents.
B. we are obligated not to harm one another.
C. we are obligated to care for strangers when it doesn't harm us to do so' (Zack 2009).

If some have misused the word democracy, that is no reason to turn against what it stands for. The democracy concept is necessary to criticize and correct abuses that falsely appropriate its name. The pragmatics of shame is limited. Public figures and others of status can be shamed for specific deeds, but there is no mechanism to shame them for the long-term benefits that they derive from the misfortunes of others because they so often mistake their status for their personal character. To evade an unpleasant decision is a moral responsibility. In making that decision, there is the further difficult question of whether something's being the lesser of two evils does make it morally permissible. The morality linked to natural disaster is a cause for anguish because choices are presented that involves loss, no matter what is done. Moreover, no two cases are likely to be exactly the same, so the kinds of choices required will need to be made anew each time. Previously accepted moral values and principles will be highly relevant to such assessment, but we cannot expect them to save us from the work of deliberation and choice. Immediate intuitions, as well as strong emotional reactions, will be relevant, but we cannot permit the unfairness of arbitrary decisions (Zack 2009).

A different type of desire to survive is observed in local communities affected by a natural disaster. It manifests itself in self-isolation from the environment during the trauma, aversion towards people visiting the area of destruction. The second group of victims—on the contrary, using their knowledge of social and administrative life, uses people who come to the disaster area (curious, tourists, photojournalists) to strengthen the message addressed to local and global authorities, to force officials to be more active and active serving to improve the situation of the local community. An example is the situation after Hurricane Katrina in 2005 (www.thesocietypages.org/socimages/2011/08/28/ disaster-tourism). Some actions taken within the same affected local community may be painful and hurtful and cause contradictory emotions.

A contradiction can also be noticed in the case of tourists' interest in natural disaster sites; when some of them decide to come to the disaster site, tourists also come to meet their own emotional and cognitive needs, which can be called natural disaster tourism or an element of dark tourism. One can see here a duality of doubts about the nature of the moral principles of visiting such places, as well as inflicting pain and economic support (leaving money spent). Among the visitors to the disaster areas are also journalists or amateur photographers, posting photos and videos on social media or the press. The question is, was this photo taken for documentation or commercial purposes? These are often comorbid goals. Moreover, what is the so-called 'A good photo' (reportage material)? In commercial terms, it is a picture that can be sold, more and more often with an image of tragedy and pain. Mental or physical images are a commodity with a specific market value. How should a photographer be assessed who takes a good photo (or film) in this sense, reaching into the intimate sphere of human suffering?

5.9 Concept of Disaster Risk Reduction: Is It a Solution to the Problems?

The Sendai Framework for Disaster Risk Reduction 2015–2030 was created during third conference in circle of the World Conference on Disaster Reduction (WCDR) that is concerned with natural disasters after the Yokohama Strategy and Plan for Action for a Safer World (1994), and the Hugo Framework for Action in Kobe in 2005. But any preventative measures should place the individual firmly at their centre. Local strategies should be ready by 2020, and risk reduction activities should take social features and the role of stakeholders into account.

A reduction in the risk of natural disasters carries a certain degree of universality which is independent of the type of natural hazard and place of occurrence, and involves three links:

1. Scientific research whose result is to create scientific bases in risk assessment, warning systems, evacuation routes, spatial development plans that take natural hazards into account, and an understanding of changes in social perception of natural hazards and adaptability.
2. Actions of the state administration, which include creating:
    - A crisis management system, implementing this and verifying this.
    - Rules for any responses from the level of institutions to individual responses.
    - A transmission of information on those states placed on alert and threatened status, such as information that would be simple to convey, be uniform, and be clearly understood by the recipient of such given information based on the state of the situation and the degree of threat.
    - A warning system that is free and available for various social groups (ethnic, religious, different status of affluence).

- Patterns and facilities for the population for preventive action.
- An informal (non-school) educational system and training to systematically complement knowledge and experience.
- Informing and educating the public about the location of those areas deemed either endangered or safe and the behaviour expected in relation to them, plus the desired response of the population to a given warning, evacuation routes, potential for changing the location of residential areas in order to avoid any confrontation with the elements, along with adaptable methods to the threat.
- Facilitating individual and group preventive action within local communities.
- Fulfilling obligations resulting from domestic and international legal arrangements.

3. Social activities that include:

- An understanding of what are natural hazards and the risks within society as a result of school education, including geographical education.
- Knowledge about the location of endangered areas and areas of safety in community, and shaping social awareness to avoid areas at risk.
- Understanding the information provided about a given situation.
- Undertaking independent and group (local) actions in the areas at risk; during any natural threats, before and after the event in accordance with the principles of ethics.
- Shaping any conscious threats to social attitudes as a readiness to take protective and corrective actions (but not omitting the solutions); and aware of the consequences for failing to take preventive measures (and so ensuring a reduction of any moral hazards).
- Mutual cooperation between management units and the local population.

The 'Triangle of Disaster Risks Reduction' (TDRR) diagram below presents society as an important link that receives information and warnings about any imminent threat, as well as those taking part in preventive activities. The TDRR includes (a) hazard and risk assessment and map creation; (b) systems' creation and modification of information, warning, risk and DRR management, and prevention; (c) society response based on knowledge and social education (formal and informal). These actions require an observance of moral principles, and any ethical elements falling within the scope of geoethics should be included in three parts of this triangle (Fig. 5.7). One area of education should also include the potential risk of losses (life, property, and social values) in various spatial scales and the principles of reducing this risk, including any ethical implications.

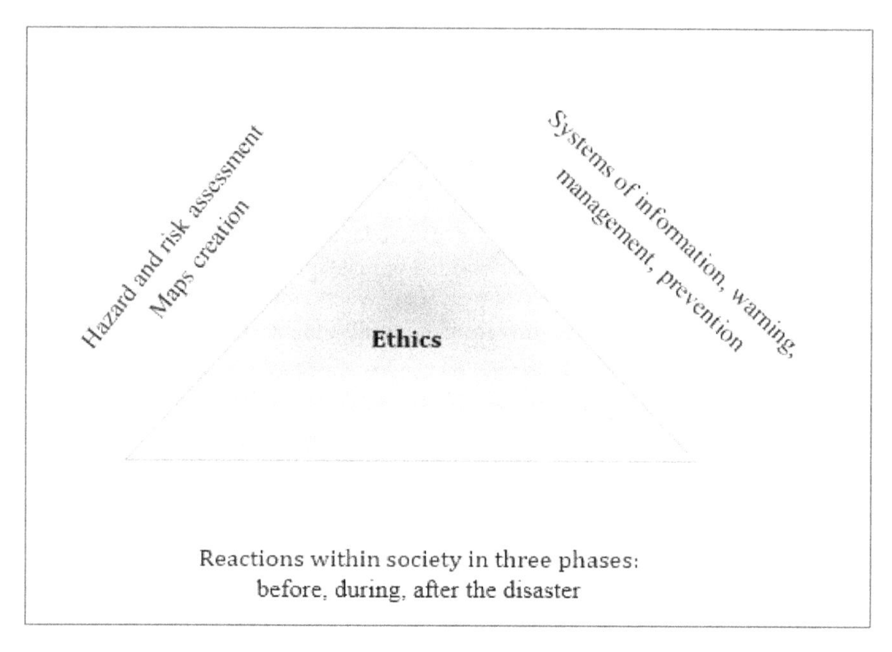

Fig. 5.7 Triangle of disaster risk reduction. *Source* Author's own elaboration.

5.10 Summary and Conclusions

The need that there is a universal model of ethical problems accompanying natural disasters has been confirmed, using many examples. This investigation shows the need to put in place rules to prevent fraud and corruption in the event of extreme natural events, threats, and natural disasters. In addition, to create a geoethical code covering nature-society issues, including risk reduction not only of geological phenomena, but also of geographic and other phenomena. A code of ethics for natural hazards and natural disasters is imperative for ethical reasons, as the disaster threatens values such as human life and wellbeing (Zack 2009), quality of life, and sustainable development. In this context, it is impossible to answer positively the question of Breczko (2017) 'are we dealing with moral progress with the development of civilization and knowledge?' Corruption scandals and unethical actions in the context of a natural disaster are the challenges of the twenty-first century.

Sigmund Freud focuses also attention to the reality and the ability to face it which is natural hazards and disasters, socioeconomic situation, and the ethics. Dietrich v. Hildebrand claims that a person's existence begins with the awakening of his moral consciousness. In the absence of it, an adult and sound mind person is always at fault (which refers to the concept of geoethics and responsibility: who has a knowledge this one is responsible). The existence or collapse of humanity depends on man's moral consciousness. Behaviour is related to motivation. When we have the ability to perceive moral aspects of reality and respond adequately to them, our moral sensitivity towards other people is mature. Moral sensitivity by Emmanuel Levinas is existence of two states of man—separation from another person and encounters with him. Therefore, moral sensitivity involves helping people in need. Giving means 'taking something out of yourself' despite your own egoism. It is also a confrontation with reality by Freud, because you cannot give up responsibility for another person. But it is related with a maturity of postmodern man. The apotheosis of freedom and equality and also individualism against the common public good and social relations are not helpful in building moral principles regarding natural threats and disasters. Speaking for geographer B.J. Berry: A socially shaped style of thinking limits an individual's perception of reality. At the end of this chapter, it is worth quoting the words and observation made at the beginning of the twentieth century by prof. Franciszek Bujak: 'The influence of natural conditions is not absolute, but depends on the character and cultural progress of people, on history, on what coexistence has done with people' (Bujak 1906). This historical sentence expressed by geographer indicate the especially importance of taking up problems arising from the relationship between nature and human, in history and today. Thus, it confirms the legitimacy of researching issues in the field of geoethics, including the issues of geography and the DRR.

As regards interpreting the term 'natural disaster', it is in relation to a situation in which we are powerless and in a set of circumstances with no obvious solution. The result is a traditional reaction—that is, a period of recovery after devastation, which is associated with the reconstruction from the resulting damage. At the same time, strong social emotions dominate any individual's attachment to their home and their desire to recreate this. The second option is to consider whether one should relocate and what preventive measures should be taken for the future or to increase spending by investing in new technological solutions, thus better protecting oneself from another disaster. This is especially the case when there is no strategic assistance from local public administration. It should be emphasized that the centre for WCDR has clearly stated that more significant funds should be directed into widespread prevention rather than towards typical reconstruction. Reconstruction should take place only in those unthreatened areas or those judged to be very low-risk, taking into account any new technological solutions that can be implemented to minimize losses (BBB concept).

A natural disaster is a powerful tool in the hands of various people (journalists, reporters, politicians, tour-operators, local communities), and pain and mental and physical condition are nowadays a specific transaction value (photo, film, report, travel to an unusual place saturated with tragedy). Natural disasters are accompanied by decisions made both at the level of government officials, large, medium, and small companies, non-profit organizations, social groups and individuals, often associated with a rational approach to a problem or the desire to survive in a position, or with the phenomenon of moral hazards. The motivation of the behaviour and the choice of decisions and actions sometimes go beyond the ethical rules. However, the evaluation of these choices, decisions, and attitudes is very complex. The existing chain of connections generates behaviours accompanying indifference, as well as actions present with sensitivity—for the good of the community.

In this context of natural disaster, both heroic help and altruism are observed, along with concentration on one's own: interests, the need to satisfy one's curiosity, ambitions, emotional, educational, and economic goals. Natural disaster is accompanied by manifestations of sensitivity and indifference. The participants of the phenomenon are not only the victims, but also groups of public officials, mass media, and tourists. Actions with specific intentions are assessed and used differently. Valuing people's behaviour in the event of a natural disaster is complex, and their attitudes are related to the purpose of the action, sensitivity, or lack thereof, but also to the knowledge and responsibility of the definition of geoethics. Elements of business ethics are linked and could be helpful to actions for DRR.

The presented context of a natural disaster seems to be important to the author from the point of view of the need to define the state of morality and shape the contemporary morality of a global and local society. Generally positive solutions commonly used—such as new technologies, increased range and quick access to places that used to be remote and inaccessible—bring with them a commercial approach to life strongly rooted in Machiavellianism. This results in the emergence of new social phenomena with the perspective of their development. At the same time, there is a need to formulate ethical principles which are a reference to the existing attitudes and behaviours towards natural threats and natural disasters. There is a need to establish moral principles deeply embedded in ethics, principles adequate and expressive to contemporary social phenomena.

Ethics should influence morality, because ethics is aimed at a thorough understanding of the basic components and principles of moral behaviour along with some, still quite general indications of how to apply these principles in the practice of life. The key question is how to relate general principles to concrete actions, which are always conditioned, by the many circumstances in which a person acts and the characteristics of his individual temperament and disposition? What are the moral standards that govern 'good' and 'bad' for DRR? So far, global conferences in Yokohama (1994), Kobe (2005), and Sendai (2015) do not raise significantly the subject of broadly understood ethics, and thus also geoethics, which deserves special attention in the context of reducing the risk of a natural disaster. Although the need to fight poverty is stressed, there are no other important indications as to the phenomena that generate them, such as corruption, methods of identification, and eradication.

Sustainable development aims at maintaining the balance and security of the natural environment and man. Sustainable development was mentioned during the world conference in Kobe. The human security aspect should also be considered in the context of natural disasters, which are the result of natural phenomena and human presence in the risk area. Sustainable development has devoted much attention in recent years. An important issue is to enable development while maintaining the values of the natural environment. It is important to preserve and create good living conditions for people in relation to the concept of 'quality of life' considering natural hazards and disasters.

Ethical issues are saddled with people forced to individual decisions in the face of danger or as a result of their function. Making a choice depends on the level of acceptability of risk, which is an individual feature or is the result of business procedures or lack thereof. The topic of acceptability of risk in the context of natural hazards has not yet been thoroughly examined and elaborated. It can be said, however, that acceptability is related to the characteristics of a given culture, as well as to individual characteristics, including empathy.

There is the need to establish rules for the prevention of fraud and corruption in the event of extreme natural hazards. Preparedness for disaster should also mean preparing for an increased number of abuses, which requires the development of new anti-corruption mechanisms.

Taking on recommendations and better integrating DRR with other process does not in itself guarantee a disaster-free world. Risks will always exist and decision-makers, necessarily, must always balance varying residual risks. It is important to acknowledge those residual risks, especially to determine if disasters could nonetheless be avoided through targeted interventions of good quality as well as right decisions that do not, in turn, cause further risk-related or other problems. If DRR is being accepted as or aimed to be a 'culture', 'contra-culture' exists, as does 'contra-development' (Lewis 2012). At the moment, the contra-culture of ethics dominates, as shown by this paper's examples connected the large scale of 'the bad and the ugly' actions (without tackling all vulnerability). Recognizing and admitting some of the drivers of vulnerability to natural hazards are a 'good' for responsibility.

The U.S. investigation has made it possible to put together a model of 'recovering from disaster', which consists of phases: loss of activity, restoration of activity in the disaster area, and two stages of reconstruction. The final phase may take preventive actions in some cases and implementing new technological solutions and thus is crucial to move people away from risk areas as suggested by the Dutch concept in programme 'Room for the Rivers', which should be treated universally in the context of risk areas. The American practice implemented in the reorganization of Rapid City, South Dakota, an area that was threatened after a catastrophic flood in 1972 shows real possibility to rationality on the development of risk areas.

It is worth emphasizing the elements of formal and informal education when it comes to reducing the risk of natural disasters. The situation definitely requires education within a wider group of society (stakeholders), apart from individuals with higher education or expert recently knowledge. Education is the basis for being receptive to information, for understanding of warning systems, and for building public awareness that is necessary for responding and implementing diversified preventive actions. Official's ethics is also necessary to create a confidence of society and accept warning system as well as use the information about natural hazards and prevention, by community.

This chapter recognizes that both the geography and geoethical approach to issues on the border of analyses of human activity and the nature include the contemporary world issues of ethics called recently geoethics which are discussed now. This chapter only presents selected examples of contemporary geoethical problems that would require further analysis and solutions. It has indicated those features determining the universality of the concept of reducing the risk of natural disasters, which may be helpful in solving the problems of loss due to natural disasters. Experiences have taught us that applying the guidelines contained in the risk reduction triangle will be useful, but only when meeting the conditions of moral principles.

So far, global conferences in Yokohama in 1994, in Kobe in 2005, and in Sendai in 2015 do not significantly raise the subject of broadly understood ethics, and thus geoethics, which deserves special attention in the context of reducing the risk of a natural disaster. Although the need to fight poverty is stressed, there are no other important signals as regards the phenomena that generate them, such as corruption, methods of identification, and eradication.

References

Abbring JH, Chiappori PA, Zavadil T (2007) Better safe than sorry? Ex Ante and Ex post moral hazard in dynamic insurance data. Tinbergen Institute Discussion Paper No. 08-075/3, Tinbergen Institute

Aghaei N, Seyedin H, Sanaeinasab H (2018) Strategies for disaster risk reduction education: a systematic review. J Edu Health Promot 7

Akkar S et al (2011) Elazıg˘ -Kovancılar (Turkey) earthquake: observations on ground motions and building damage. Seismolog Res Lett 82(1):42–58. https://doi.org/10.1785/gssrl.82.1.42

Alexander D (2017) Oxford research encyclopedias. Natural Hazards Science. Corruption and the Governance of Disaster Risk. Subject: Resilience, Risk Management, Earthquakes, Policy and Governance, Cultural Perspectives. available from http://naturalhazardscience.oxfordre.com (13.09.2018).

Allan M (2015) Geotourism: an opportunity to enhance geoethics and boost geoheritage appreciation. Geol Soc London Spec Publ 419:25–29

Ambraseys N, Bilham R (2011) Corruption kills. Nature 469:153–155

Anderson H, James and Cheryl Gray W (2007) Policies and Corruption Outcomes, 'Anticorruption in Transition 3. Who is Succeeding...and Why? The World Bank, pp 43–77

Asheim BT (2020) Economic geography as regional contexts' reconsidered—implications for disciplinary division of labour, research focus and societal relevance. Norsk Geografisk Tidsskrift—Norwegian Journal of Geography 74(1):25–34. https://doi.org/10.1080/00291951.2020.1732457

Baker T (1996) On the Genealogy of Moral Hazard, Faculty Scholarship. Paper 872. Univ Pennsylvania 75(2):237–292

Bales K (2007) What predicts human trafficking? Int J Comparat Appl Crim Just 31(2):269–279

Birkeland IJ (1998) Nature and the 'cultural turn' in human geography. Norsk Geografisk Tidsskrift—Norwegian J Geograph 52(4):221–228

Bobrowsky P, Cronin VS, Di Capua, G, Kieffer, SW, Peppoloni S (2017) The emerging field of geoethics. In: Gundersen LC (eds) Scientific integrity and ethics with applications to the geosciences, Chapter 6. Special Publication American Geophysical Union, John Wiley and Sons, Inc.

Borowiec M (2006) 'Etyczne aspekty globalizacji w procesie kształtowania przedsiębiorczości' ['Ethical aspects of globalization in the process of shaping entrepreneurship'] *Przedsiębiorczość—Edukacja* 2:185-192. Available from www.cejsh.icm.edu.pl. Accessed on 15 Sept 2018

Branigan T (2011) China launches corruption inquiry into railway minister. The Guardian 13 February (10.09.2015).

Breau S, Carr I (2009) Humanitarian aid and corruption. Working paper. Retrieved from https://papers.ssrn.com/sol3/papers.cfm?abstract_id=1481662

Breczko J (2017) *Kilka zastosowań empatii w kwestiach filozoficznych*, wydawnictwo, (ed. red.) J. Mysona Byrska, J. Synowiec., WN UPJPII Wyd. Naukowe, Kraków 9:5–40
Breit E, Lennerfors TT, Olaison L (2015) Critiquing corruption: A turn to theory. Ephemera. Theory and Politics in Organisation. ISSN 1473-2866 (Online) ISSN 2052-1499 (Print) www.ephemerajournal.org volume 15(2):319-336.
Britannica (2021) www.britannica.com
Bujak F (1906) Historia stosunków gospodarczych. Ekonomista" t. 2
Byrska-Rąpała A (2008) 'Geoetyka a społeczna odpowiedzialność przemysłu surowców energetycznych' ['The Concept of Social Responsibility in the Energy Sector and the Principles of Geoethics in Building Enterprise Value] Gospodarka surowcami energetycznym 24(4):41–52
Ceccarelli G (2001) Risky business: theological and canonical thought on insurance from the thirteenth to the seventeenth century. J Med Early Mod Stud 31:607–658
Corunt P, Swyngedouw E (2000) Approaching the society-nature dialectic: a plea for a geographical study of the environment. Belgeo 1-2-3-4:37–46
Crescimbene M, La Longa F, Camassi R, Pino NA (2015) The seismic risk perception questionnaire. Geol Soc London Spec Publ 419:69–77
Crosweller M (2022) Disaster management and the need for a relational leadership framework founded upon compassion, care, and justice. Climate Risk Manag 35(2022):100404
Datta PS (2015) Ethics to protect groundwater from depletion in India. Geol Soc London Spec Publ 419:19–24
De Pascale F, Bernardo M, Muto F, Tripodi V (2015) Geoethics and seismic risk perception: the case of the Pollino area, Calabria, southern Italy and comparison with communities of the past. Geologic Soc London, Spec Publ 419:87–102
De Rubeis V, Sbarra P, Sebaste B, Tosi P (2015) Earthquake ethics through scientific knowledge, historical memory and societal awareness: the experience of direct internet information. Geol Soc London Spec Publ 419:103–110
Dembe AE, Boden LI (2000) Moral hazard: question of morality? New Sol 10(3):257–279
Disaster Fraud Task Force, DFTF (2011) Report to the attorney general for fiscal year 2011. U.S. Dep of Justice, available from www.justice.gov (19.09.2018).
Dumanowski B (1981) 'Geografia regionalna jako dyscyplina badawcza' ['Regional geography as a research discipline'] Przegląd Geograficzny 53(1):87–94
Dziadkiewicz A (2012) Znaczenie etyki biznesu w kształtowaniu współczesnych norm i wartości. 'Rola Przedsiębiorczości w edukacji', Przedsiębiorczość—Edukacja, 8:155–168
Engstfeld A, Killebrew K, Scott C, Wiser J, Freitag B, El-Anwar O (2010) Tsunami safe haven project—report for Long Beach, Washington. Department of urban design and planning, college of built environments, University of Washington
Escaleras M, Anbarci N, Register CA (2007) Public sector corruption and major earthquakes: a potentially deadly interaction. Public Choice 132:209–230
Estrada MAR, Staniewski, MW, Ndoma I (2018) Evaluating corruption under the application of the SOCIO-economic development desgrowth index (Đ-index): the cae of Guatemala. Qual Quant Int J Methodol 52(3):1137–1157. https://doi.org/10.1007/s11135-017-0508-5
Ferreira C, Engelschalk M, Mayville W (2007) The challenge of combating corruption in customs administrations. In: Edgardo Campos J, Sanjav P (eds) The many faces of corruption: tracking vulnerabilities at the sector level, pp 367–386
Ferrero E, Magagna A (2015) Natural hazards and geological heritage in Earth science education projects. Geol Soc London Spec Publ 419:149–160
Foltz RC (2002) Iran's water crisis: cultural, political, and ethical dimensions. J Agricult Environ Ethics 15:357–380
Fox.news (2017). https://www.foxnews.com/politics/chris-christie-harvey-aid-should-not-be-held-political-hostage-like-sandy-aid
Friedman HS (2020) Ultimate price: the value we place on life. University of California Press
Glanz MH, Zonn IS (2005) The Aral sea: water, climate, and environmental change in Central Asia. The World Meteorological Organization. Weather—Climate. Water. WMO-No.982. Geneva

Golden MA, Picci A (2005) Proposal for a new measure of corruption, illustrated with Italian data. Econ Polt 17(1):954–1985

Green P, Ward T (2004) State Crime: Governments. Violence and Corruption, Pluto Press, London

Grundy P (2008) Disaster reduction on coasts. In: Proc., IStructE Centenary Conference, Hong Kong, January, pp 246–263

Gryżenia K (2009) Współczesna moralność bez etyki? 'Annal Ethics Econ Life 12(1):205–217 Archidiecezjalne Wydawnictwo Łódzkie, www.annalesonline.uni.lodz.pl/archiwum/2009/2009_01_gryzenia_205_217.pdf

Hadžialić S, Phuong VT (2020) Media ethics within the fake news challenges during the covid-19 pandemic. Studia i Analizy Nauk Politycznych 2:33–46

Hanson S, Vitek JD, Hanson PO (1979) Natural disaster: long-range impact on human response to future disaster threats. Environ Behav 14:171–185

Hill L (2012) Ideas of corruption in the Eighteenth Century: The competing conceptions of Adam Ferguson and Adam Smith, Corruption. In: Barcham M, Hindess B, Larmour P (eds) ANU Expanding the docus, E Press

Hoogvelt AMM (1976) The sociology of developing societies. Macmillan, London

International Press Foundation, IPF (2015) Corruption in disaster recovery: disturbing tales from the Philippines and Haiti, by Makoi Popioco, 27 Nov 2015. Available from www.the-ipf.com (19.09.2018)

Ippolito F (1968) La natura e la storia. Vanni Scheiwiller. Milano

Johnston M (2009) Poverty and corruption. Forbes Magazine 22 January (14.04.2011)

Kałuski S (2017) Blizny historii. Geografia granic politycznych współczesnego świata [Scars of history. Geography of the political boundaries of the modern world.] Wydawnictwo Akademickie Dialog. Warszawa.

Kantowicz E, Skotnicki, M. (1984). Trends in regional geography. Miscellanea Geographica, 1(1):153–162

Kantowicz E, Roge-Wisniewska M (2012) Cywilizacja a Środowisko. Wyzwania i dylematy [Civilization and Natural Environment—Challenges and Dilemmas]. Wydawnictwo WGiSR UW, Warszawa

Kates RW (1971) Natural hazard in human ecological perspective: hypotheses and models. Econ Geograph 47(3):438–451

Kesteloot C, Bagnoli L (2021) Human and physical geography: can we learn something from the history of their relations?, Belgeo [Online], 4 | 2021, Online since 02 March 2022, connection on 12 September 2022; https://doi.org/10.4000/belgeo.52627.

Kundzewicz ZW, Kowalczak P (2008) Zmiany klimatu i ich skutki [The Climate Change and the Effects], 1st rev. ed. Kurpisz S.A. Poznań

Kuroiwa J (1982) Studies on the prevention of earthquake disasters and their application in urban planning in Peru Mimeo: Workshop on Planning for Human Settlements in Disaster- prone Areas, UNCHS 1983 Nairobi

Kuroiwa J (1986) Planning and management of regional development for earthquake disaster mitigation. Mimeo UNCRD

Kuroiwa J (2002) Communication ISDR Earth Summit Debate http://www.earthsummit2002.org/debate

Kushida KE (2012) Japan's Fukushima nuclear disaster: Narrative, analysis and recommendations. Working paper Stanford, CA: Walter H. Shorenstein Asia-Pacific Research Center, Stanford University

Kuzior P (2016) Business ethics and sustainable development interdisciplinary theoretical and empirical studies. No. 2. Theoretical background of practical operations within CSR and Sustainable Development, Silesian Center for Business Ethics and Sustainable Development. Zabrze

Lewicka-Strzałecka A (2018) Pomiar korupcji i jego ograniczenia [Corruption Measures and Their Limitations], Annales. Etyka w Życiu Gospodarczym / Annales. Ethics in Economic Life 21(1):7–20 https://doi.org/10.18778/1899-2226.21.1.01

Lewis J (2003) Housing construction in earthquake-prone places: perspectives, priorities and projections for development. Austr J Emerg Manag 18(2)
Lewis J (2008) The worm in the bud: Corruption, construction and catastrophe. In: Lee B (ed) Hazards and the built environment: attaining built-in resilience, Routledge, pp 238–263
Lewis J (2012) The good, the bad and the ugly: disaster risk reduction (DRR) Versus Disaster Risk Creation (DRC). PLOS Currents Disasters. 2012 Jun 21. Edition 1. https://doi.org/10.1371/4f8d4eaec6af8
Limaye SD (2015) A geoethical approach to industrial development of georesources and groundwater use: 13 the Indian experience. In: Peppoloni S, Di Capua G (eds) Geoethics: the role and responsibility of geoscientists, pp 13–18
Lollino G, Arattano M, Giardino M, Oliveira R, Peppoloni S (2014) Engineering geology for society and territory—volume 7. Education, Professional Ethics and Public Recognition of Engineering Geology. Set: Engineering Geology for Society and Territory 2014, IAEG XVII Congress, Springer
Martin R, Sunley P (2022) Making history matter more in evolutionary economic geography. ZFW—Adv Econ Geograph 66(2):65–80
Mausolf N (2011) Die Stalinbahn—Trilogie. Auf Spurensuche am Polarkreis. Books on Demand GmbH, Norderstedt
McCullough D (2001) The path between the seas: the creation of the Panama Canal, 1870–1914. Simon and Schuster, New York, London, Toront, Sydney, Singapure
Meriläinen E, Mäkinen J, Solitander N (2020) Blurred responsibilities of disaster governance: the American Red Cross in the US and Haiti. Pol Govern 8(4):331–342
Min J, Birendra KC, Seungman K, Lee J (2020) The impact of disasters on a heritage tourist destination: a case study of Nepal earthquakes. Sustainability 12:6115. https://doi.org/10.3390/su12156115
Natural hazards Hazards Observer (2002) WMO
Nemec V (1992) Ethical geology in the education process. In: 29th international geological congress, Kyoto, Japan. 24 August–3 September 1992. Section II-24–1 "New ideas and techniques in geological education", vol 3(3)
NIK Report 2014 after the flood in 2014. http://www.gazetaprawna.pl/artykuly/787478,raport-nik-dzialania-przeciwpowodziowe-samorzadow.html (21.10.2015).
Nikitina NK (2016) Geoethics: theory, principles, problems. Monograph. 2nd edition, revised and supplemented. Moscov: Geoinformmark, Ltd
OECD Glossries (2008) Corruption. A glossary of international standards in criminal law, OECD, pp 22, 23. www.oecd.org/daf/nocorruption/
Ogen O (2008) The economic lifeline of British global empire: a reconsideration of the historical dynamics of the Suez Canal, 1869–1956. J Int Soc Res 1(5):523–533
Ozerdem A (1999) Tiles, taps and earthquake-proofing: lessons for disaster management in Turkey. Environ Urbaniz. https://doi.org/10.1177/095624789901100215
Palm R (1981) Public response to eartquake hazard information. Annal Assoxiation Am Geograph 71:389–399
Pauly MV (1968) The economics of moral hazard: comment. Am Econ Rev 58:531–537
Peppoloni S (2012) Social aspects of the earth sciences. interview with Prof. Franco Ferrarotti. Annal Geophy 55(3):347–348
Peppoloni S, Di Capua G (2012) Geoethics and geological culture: awareness, responsibility and challenges. Ann Geophy 55(3):335–341. https://doi.org/10.4401/ag-6099availablefromwww.geoethics.org(5.07.2018)
Peppoloni S, Di Capua G (2015) Geoethics: the role and responsibility of geoscientists. Geological Society, London, Special Publications. The Geological Society. Available from www.geoethics.org. Accessed on 5 July 2018
Peppoloni S, Di Capua G, Bobrowsky P, Cronin V (2017) Emerging field of geoethics. In: Gundersen LC (ed) Scientific integrity and ethics with applications to the geosciences. Special Publication American Geophysical Union, John Wiley and Sons

Pocock DCD (1974) The nature of environmental perception. Univ. of Durham
Potter VR (1971) Bioethics: bridge to the future. Prentice-Hall, Englewood Cliffs, NJ
Potter VR (1988) Global bioethics. Michigan State University Press, Building on the Leopold Legacy
Report (2011) DFTF, Disaster fraud task force. U.S. Dep. of Justice. FEMA
Ripp M, Rodwell D (2015) The geography of urban heritage. Hist Environ Policy Pract 6(3):240–276
Rowell D, Luke B (2012) A history of term 'Moral Hazard.' J Risk Insuran 79(4):1051–1075. https://doi.org/10.1111/j.1539-6975.2011.01448.x
Rucińska D, Walewski A (2004) 'Kształcenie w Zakładzie Geografii Regionalnej' ['Education at the Regional Geography Department']. Prace i Studia Geograficzne 34:57–67
Rucińska D (2005) *Definicje przyrodniczych zjawisk ekstremalnych w podręcznikach szkolnych na poziomie ponadgimnazjalnym.* In Mat. konf.: XXXI Ogólnopolski Zjazd Agrometeorologów, Bydgoszcz, pp 14–16
Rucińska D (2011) Social education on extreme natural events in view of extreme floods and landslides in Poland. Prace i Studia Geograficzne 48:173–185
Rucińska D (2012) Ekstremalne zjawiska przyrodnicze a świadomość społeczna [Extreme natural phenomena and social awareness]. Uniwersytet Warszawski Wydział Geografii i Studiów Regionalnych
Rucińska D (2012) Ekstremalne zjawiska przyrodnicze a świadomość społeczna, Wydawnictwo Wydziału Geografii i Studiów Regionalnych UW
Rucińska D, Lechowicz M (2014) Natural hazards and disaster toursim. Miscellanea Geographica—Reg Stud Develop 18(1):17–25. https://doi.org/10.2478/mgrsd-2014-0002
Rucińska D (2014) Interdyscyplinarność i uniwersalność koncepcji redukcji ryzyka klęsk żywiołowych [Interdisciplinarity and the versatility of the concept of disaster risk reduction]. Prace i Studia Geograficzne 55:119–131
Rucińska D (2016) Natural disaster tourism as a type of dark tourism. Int J Soc Behav Educ Econ Bus Industr Eng World Acad Sci Eng Technol 10(5):1276–1280
Rucińska D (2017) Klęska żywiołowa—przejawy wrażliwości i obojętności' ['Natural disaster—manifestations of sensitivity and indifference']. In: Mysona Byrska J, Synowiec J (eds) Obojętność i wrażliwość a życie publiczne. Monografia wieloautorska. Indifference and sensitivity and public life. Multi-author monograph, vol 9, pp 1–15. WN UPJPII, Kraków
Rucińska D (2021) Redefinicja klęski żywiołowej i elementów ryzyka strat. In: Szlahetko JH, Bochentyn A (eds) Redefinicja klęski żywiołowej i elementów ryzyka strat. Konceptualizacja i dyskusja. Instytut Metropolitalny. Bernardinum, Pelpin, 13–34
Schweitzer A, Naish JP (1923) Manuscript civilization and ethics. A. and C. Black, London
Silva E, Sá AA, Roxo MJ (2015) From planet Earth to society: a new dynamic in Portugal concerning geoscience education and outreach activities. Geolog Soc London Spec Publ 419:141–147
Sobala E (2015) Prowizorka na lata. Powodzianie z 1997 roku w 60-letnich barakach http://wroclaw.tvp.pl/19245564/prowizorka-na-lata-powodzianie-z-1997-roku-w-60letnich-barakach (13.12.2015).
Sobel RS, Leeson PT (2006) Government's response to Hurricane Katrina: a public choice analysis. Public Choice 127:55–73
Sorensen J (2014) Why are natural disasters breeding grounds for corruption?, March 3, 2014, This story originally appeared on Talkingpointsmemo.com on March 3, By Juliet Sorensen
Sroka R (2017) *Zasada niepowodowania cierpienia w etyce biznesu,* [w:] J. Myson Byrska. J. Synowiec, Obojętność i wrażliwość w życiu publicznym, Wydawnictwo Naukowe, UPJPII, Kraków 9:233–252
Stefanile S (2020) "Evento virtuale su "Corruption in the time of COVID-19: a double-threat". Intervento dell'Ambasciatore Stefano Stefanile, Vice Rappresentante Permanente dell'Italia presso le Nazioni Unite, " https://italyun.esteri.it/rappresentanza_onu/it/comunicazione/archivio-news/2020/12/lancio-virtuale-del-group-of-friends.html (09.02.2021).
Stone PR (2006) A dark tourism spectrum: towards a typology of death and macabre related tourist sites, attractions and exhibitions. Tour Interdisciplin Int J 54(2):145–160

Storpert M (2011) Why do regions develop and change? the challenge for geography and economics. J. Econ. Geograp. 11(2011):333–346
Stryjakiewicz T (2022) Emergence of socio-economic geography and spatial management as a scientific discipline in the new classification of science in Poland. Quaestiones Geographicae Bogucki Wydawnictwo Naukowe, Poznań 40(4):7–14. https://doi.org/10.2478/quageo-2021-0033
Sugimoto M, Iemura H, Shaw R (2010) Tsunami height poles and disaster awareness. Memory, education and awareness of disaster on the reconstruction for resilient city in Banda Aceh, Indonesia. Disaster Prevent Manage 19(5):527–540
Sullivan HR (2019) Voluntourism. AMA J Ethics 21(9):815–822. https://doi.org/10.1001/amajethics.2019.815
Tait R (2006) After the earthquake, Bam battles with heroin and AIDS. The Guardian (11.05.2006). talkingpointsmemo.com, IPF
Tanaś T (2006) Tanatoturystyka—kontrowersyjne oblicze turystyki kulturowej. Peregrinus Cracoviensis 17:85–100
Tanner A, Árvai J (2018) Perceptions of risk and vulnerability following exposure to a major natural disaster: the calgary flood of 2013. Risk Anal 38(3):548–561
Taylor PJ, O'Keefe P (2021) In praise of Geography as a field of study for the climate emergency. Geograph J 187:1–8. https://doi.org/10.1111/geoj.12404, https://doi.org/10.1111/geoj.12404
The Telegraph (2020). www.telegraph.co.uk/travel/comment/wuhan-dark-tourism-coronavirus/ (12.02.2021)
Tomazos K, Cooper W (2011) Volunteer tourism: at the crossroads of commercialisation and service? Curr Issues Tour https://doi.org/10.1080/13683500.2011.605112
Treder M (2006) About geoethical nanotech. www.crnano.typepad.com. Accessed on 17 July 2006
UNICEF (2012) Disaster risk reduction and education outcomes for children as a result of DRR activities supported by the EEPCT programme. United Nations Children's Fund Education Section, Programme Division Three United Nations Plaza New York
Usuzawa M, Telan EO, Kawano R, Dizon CS, Alisjahbana B, Ashino Y, Egawa S, Fukumoto M, Izumi T, Ono Y, Hattori T (2014) Awareness of disaster reduction frameworks and risk perception of natural disaster: a questionnaire survey among philippine and indonesian health care personnel and public health students. Tohoku J Exp Med 233(1):43–48
Vanem E (2012) Ethics and fundamental principles of risk acceptance criteria. Safety Sci 50:958–967
Varian HR (2010) Intermediate mircoeconomics: a modern approach, 8th edn. W.W. Norton and Company, New York
Vecchia P (2015) The oil and gas industry: the challenge for proper dissemination of knowledge. In: Peppoloni Di Capua G (eds) Geoethics: the role and responsibility of geoscientists, pp 133–140
Walewski A, Kantowicz E (2010) The relations between man and the natural environment as the methodological basis for delimitation of regions. Miscellanea Geographica 14:295–301
Walker P (2005) Opportunities for corruption in a celebrity disaster. Background paper for the expert meeting on corruption prevention in tsunami relief, Jakarta' April 7–8, 2005, ADB/OECD Anti-Corruption Initiative for Asia and the Pacific. Transparency International, Berlin
Walker GR, Grundy P, Musulin R (2011) Disaster risk reduction and wind
Wamsler C, Brink E, Rantala O (2012) 'Climate change, adaptation, and formal education: the role of schooling for increasing societies' adaptive capacities in El Salvador and Brazil". Ecol Soc 17(2):1–19
WCO Research Paper No. 7 (2010) Overview of Literature on Corruption, 2010
Weckroth M, Ala-Mantila S (2022) Socioeconomic geography of climate change views in Europe. Global Environ Change 72:102453
Whatmore S (2014) Nature and human geography. In: Cloce P, Crang P, Goodwin M (eds) Introducing human geographies. Routledge, London-New York, pp 152–162
White GF (1974) Natural hazards research: concepts, methods, and policy implications. In: White GF (ed) Natural Hazards: Local, National, Global. Oxford University Press, New York, pp 3–16

Winter RA (2000) Optimal insurance under moral hazard. In: Dionne G (eds) Handbook of insurance. Huebner international series on risk, insurance, and economic security, vol 22. Springer, Dordrecht. https://doi.org/10.1007/978-94-010-0642-2_6.
Wites T (2008) Forms of and prospects for the development of gulag tourism in Russia', tourism in the New Eastern Europe. global challenges—regional answers. In: International conference, Warsaw. Available from www.depot.ceon.pl (20.09.2018)
Wood N, Jones J, Schelling J, Schmidtlein M (2014) Tsunami vertical-evacuation planning in the U.S. Pacific Northwest as a geospatial, multi-criteria decision problem. Int J Disaster Risk Reduction 9:68–83
Wróblewski D (2014) Przegląd wybranych dokumentów normatywnych z zakresu zarządzania kryzysowego i zarządzania ryzykiem wraz z leksykonem. Wydawnictwo CNBOP-PIB. Józefów
Wyss M, Peppoloni S (2014) Geoethics, ethical challenges and case studies in earth sciences. Elsevier, Amsterdam-Osford-Waltham
Yamamura E (2014) Impact of natural disaster on public sector corruption. Public Choice. https://doi.org/10.1007/s11127-014-0154-6
Yamamura E (2013) Public sector corruption and the probability of technological disasters, EERI Research Paper Series, No. 02/2013, Economics and Econometrics Research Institute (EERI), Brussels http://hdl.handle.net/10419/142654.
Zack N (2009) Ethics for disaster. Published in the United States of America by Rowman and Littlefield Publishers, Inc.
Żardecka M (2017) Wrażliwość jako zdolność dostrzegania moralnych aspektów rzeczywistości. [w:] J.Myson Byrska J, Synowiec (eds) Obojętność i wrażliwość w życiu publicznym, Wydawnictwo Naukowe, UPJPII, Kraków, 9, 41–60
Ziervogel G, Pelling M, Cartwright A, Chu E, Deshpande T, Harris L, Zweig P (2017) Inserting rights and justice into urban resilience: a focus on everyday risk. Environ Urbanizat 29(1):123–138

Chapter 6
Summary and Conclusions

Tomasz Wites

Abstract The reality never ceases to surprise us and add new unknown chapters to life. This book tries to follow the challenges and circumstances faced by the world community in the early 2000s. The publication is presenting the issue of global problems at a comparative level and in a comprehensive way. Up until now, when issues of that nature were tackled, only selected aspects of environmental and socio-economic issues were covered, whereas the global part of deliberations served only as an introduction to contributing analyses. The premise for the creation of this publication was measures designed to determine the meaning of geography and the reinforcement of its position among other disciplines that take up different global problems. The synthetic contribution of this publication to expanding knowledge about global issues is as follows: an analysis of selected problems related to the natural environment, economic, social and ethical conditions of the modern world was made. The main thesis of the publication advocates taking into account the spatial aspects in studies on selected global issues. The adopted narrative is a geographical perspective that allows to take into account the spatial element in explaining the mechanisms of processes related to natural, political, ethical, social and economic issues.

Civilizational achievements create opportunities for humans and make life easier. At the same time, the transformation of the world goes hand in hand with negative environmental and social phenomena, among them those already known in the past and new ones, too, which emerge in an increasingly automated reality. They are characterised by variable conditions of establishment and varied forms. Many elements of global issues in the contemporary world are hardly quantifiable, are subject to a selective mode of registration, whereas the figures reflect only a selected range of behaviour, not the actual distribution thereof. Equally important is the transparency of data—which in many cases is formally limited. It is geographical perspective that

T. Wites (✉)
Faculty of Geography and Regional Studies, University of Warsaw, Krakowskie Przedmieście Str. 30, 00-927 Warsaw, Poland
e-mail: t.wites@uw.edu.pl

K. Podhorodecka and T. Wites (eds.), *Global Challenges*, Springer Geography,
https://doi.org/10.1007/978-3-031-60238-2_6

is of key importance for defining the research problem identified in this publication. The meaning of this term is, at the same time, comprehensive and fuzzy.

The reality never ceases to surprise us and add new unknown chapters to reality. This book tries to follow the challenges and circumstances faced by the world community in the early 2000s. The publication is presenting the issue of global problems at a comparative level and in a comprehensive way. Up until now, when issues of that nature were tackled, only selected aspects of environmental and socio-economic issues were covered, whereas the global part of deliberations served only as an introduction to contributing analyses. The premise for the creation of this publication were measures designed to determine the meaning of geography and the reinforcement of its position among other disciplines that take up different global problems. The synthetic contribution of this publication to expanding knowledge about global issues is as follows: an analysis of selected problems related to the natural environment, economic, social and ethical conditions of the modern world was made.

The main thesis of the publication advocates taking into account the spatial aspect in studies on selected global issues. The adopted narrative is a geographical perspective that allows to take into account the spatial element in explaining the mechanisms of processes related to natural, political, ethical, social and economic issues.

The axis for this perspective should contain the identification, description and explanation of the location of particular environmental, socio-economic and ethics phenomena. The process for each stage should take into account limitations and propose ways to overcome these. Spatial studies embody in the particularity of research varied depending on the scale of the study—from a global scale, through regional up to local. Analysis may be carried out in units of territorial and administrative division or other spatial separations that do not require institutional aggregation.

The ordering of issues in the axis from environmental to socio-economic topics, taking into account the ethical dimension of the perception of reality, creates a clear axis for the discussed considerations. The complicated nature of human behaviour can be analysed using an approach characteristic both for a scientific and practical orientation. Global problems are, above all, the domain of approach, which explaining environmental and social processes and phenomena in space around us by means of studying the notions and behaviour of man. Due to the diversity and multifaceted nature of the addressed issue, it was decided to take into consideration four research approaches.

The first way to delineate a geographical perspective in studies on global problems is the environmental approach. Listed among the environmental factors that potentially affect the extent of global problems are: loss and fragmentation of habitats, and the biodiversity loss. Perceived in zonal or azonal terms, the differentiation of features of the environment may correlate to mechanisms of impact on global problems phenomena. The factors determining the development of global problems, and tied to explanations of these issues, include: loss and fragmentation of habitats and the biodiversity loss. Isolated areas were selected for the case studies—the Galapagos Islands, Mauritius, and Madagascar. Additionally, the isolation index was characterized, and its potential for application is shown in determining the level of isolation of islands in relation to the threat of colonization by individual invasive species of plants and animals.

The fact that habitats are shrinking is a serious problem for biodiversity; nevertheless, another type of threat is the destruction of habitats in the form of their fragmentation. Important environmental problem contributing to the reduction of biodiversity is the spread of invasive alien species (IAS), non-native species that are alien to a given ecosystem. Selected examples of island areas that have suffered from loss of habitats and species have shown a wide range of environmental problems occurring in different parts of the world. Another important problem analysed in this book was international trade in endangered species, for example the illicit trade in rhinoceros horn and ivory smuggling. The illustrated cases give an important answer in terms of the mistakes made and give ways to solve a complicated situation. Harms which influenced to the environment and to the culture or spatial development were illustrated.

Another aspect discussed in the chapter was the sustainable development concept and its application in the area of the Alps. These goals were achieved in the theoretical and empirical parts of the work. The theoretical part explains the issues of sustainable tourism, which, although relatively new. Statistical data from official, verified sources were used, and strategic documents at the international and national level were analysed. In the empirical part, examples of activities from the described region that are consistent with the principles of sustainable tourism development were examined. The brace of the chapter was a list of negative and harmful disadvantages such as: loss of biodiversity, littering by tourists, low-status jobs, creating a tourist economic monoculture or inflation pressure.

The second way to delineate a geographical perspective in studies on global problems is the social approach. Among demographic conditions that determine the decline or development of global problems are numerous topical threads that are tied to the general rise of the population in the world (overpopulation, acceptance or stigmatising of birth control, the atrophy of polygamy, progressing migration). Yet another theme category related to demography is the population ageing, including the acceptance or disapproval thereof, leading to the development of negative social behaviour. Blue zones, where a combination of life style elements, diet and philosophy allow inhabitants to reach longevity with the life quality being very diverse, constitute a research challenge. Moreover, a wide range of possible cognitive topic is connected to migration processes. The contextual and processual nature of migration were highlighted. It pointed out five aspects, which are connected with economic, social, political, technological transformation, and environmental changes, and tried to demonstrate that these factors create interconnected conditions, which are frames for the complex, variable international migration movements. The publication gives the reader an opportunity to find answers to the following questions: do we live in an age of unprecedented migration, how have the migration situation changed in Europe over the last 60 years or how big is the influence of COVID-19 pandemic on the international migration. The contextual and processual nature of migration were highlighted.

The superior objective of the research is optimal execution of set tasks that can be used by public administration or private sector workers. The aspects of putting knowledge from the research into practice, as proposed in the dissertation, is considered an argument in favour of including the knowledge in spatial planning decisions. The postulate of practicality was underscored in papers of advocates of various theoretical and methodological approaches in socio-economic geography, which recognizes as equal actions and practice.

We live in a reality of constant changes taking place in the social environment. Human influence on the functioning of natural processes taking place on Earth is constantly increasing, and its effects are noticed at various spatial scales. The universality, comprehensiveness and practicality of social research allow for monitoring and explaining the changes taking place in local, regional and global terms.

The third way to delineate a geographical perspective in studies on global problems is the economic approach. Listed among the factors that potentially affect the extent of global problems are: global economic crisis and economic impact of COVID-19 pandemic on global economic growth or the economic problems are related to the income inequalities and the escape of wealthy people to tax haven. In addition, cyclical economic crises hit the poor more often than healthy people. In the group of economic conditions, issues of progress, growth and economic development are among the themes that may potentially align with interests in global problems. Yet another issue of interest are mechanisms that determine the level of economic openness and of correlation to environmental, socio-economic and ethics issues. Economic inequalities, determining uneven access to various goods, are closely tied to social problems.

This context presents the impact of the global economic crisis on the tourism sector in the world. It highlights which tourist destinations suffered the most during the global economic crisis and which areas did not emerge for years from the recession associated with the global economic crisis. In addition, the conditions affecting the reaction of the tourist sector to the global economic crisis are also discussed, as are the dependent countries and territories that, paradoxically, recorded an upward trend in incoming tourism during the crisis. Content on the subject provides important information about economic impact of COVID-19 pandemic on the pace of global economic growth and the global tourism sector or a blacklist of tax havens and the possibility of preventing their creation. The global of economic crisis on tourism sector was shown. The impact COVID-19 pandemis as crisis situation for tourism sector was more huge (decreased by 70–80% in foreign tourists arrivals), but to fully assest it we need to wait several years to get the scrutiny statistics.

The forth way to delineate a geographical perspective in studies on global problems is the political approach selected political problems—decommunising public space. The adopted Central European perspective made it possible to illustrate iconoclasm in respect of Warsaw monuments from the 1945–1989 period and post-soviet monuments in Ukrainian cities. As highlighted in chapter, Warsaw's space (including that of a symbolic nature) thus represents a field in which clashes have taken place, and there are demonstrations of different attitudes to Poland's historical past to be seen, as well as divergent political views. The ultimate persisting manifestation of this is the co-occurrence in the cityspace of monuments of content relating to the former system, as well as others of a typically anti-Communist nature.

The book also discusses social problems related to the contemporary world, as well as ethical problems, which mainly include: natural hazards and disasters, concept of disaster risk reduction, acceptance of risk, corruption and poverty connected with natural hazards and disasters. Another major issue, mentioned above, is the phenomenon of moral hazard, which relates to responsibility. The phenomenon of moral hazard may be feared in various ways, i.e., due to the citizen's lack of a sense of duty and to take any action in connection with preparation for natural hazards and activities aimed at reducing their socio-economic vulnerability to natural hazards. One of the wider problems is corruption. The possibility of corruption developing during a natural disaster as well as the COVID-19 pandemic was emphasized. This chapter recognizes that both the geography and geoethical approach to issues on the border of analyses of human activity and the nature include the contemporary world issues of ethics called recently geoethics which are discussed now.

One of the fruitful conclusion in this part of the publication is that risks will always exist and decision-makers, necessarily, must always balance varying residual risks. It is important to acknowledge those residual risks, especially to determine if disasters could nonetheless be avoided through targeted interventions of good quality as well as right decisions that do not, in turn, cause further risk-related or other problems.

All of the goals stipulated in this publication have been achieved, while all these have been justified and proven. Research on global problems in the contemporary world was realised in three aspects: environmental, socio-economic and ethical. The role of geography in each of these approaches is defined differently, which translates to the position of that discipline among other fields that tackle global problems issues. The interdisciplinary approach is optimal for the analysis of global problems. The subject is universal in its application and allows thus for a reference of the outcome to projects carried out in many corners of the world. Geographical identity defines a methodological workshop, specific to the various levels of reference and the applicability of solutions regarding global problems, tailored to various spatial scales.

The approach of geographers to presenting and understanding of global problems is the search for adequate ways of description, of explanation of phenomena and processes, and of discovery of regularities. The theoretical and cognitive interest in the nature of environmental and social order combines with the practical use of results obtained. These may serve to reduce or limit action considered harmful for individuals, social groups or the whole society.

The outcome of these theoretic and methodological contemplations is a proposal for a questions, on the other—to indicate feasible solutions. In the case of global problems, the goal of this publication is to suggest a set of solutions corresponding to the assumptions of practicality of geography. The postulate of professionalism of geography worded by that scholar indicates that geography should not limit itself to the description, explanation of phenomena and processes, nor to the discovery of regular features. Geography should, in fact, formulate opinions that may serve as a foundation for spatial planning decisions.

What is important in geographical studies is the regional perspective which serves to denote the conditions affecting the cause and the process of changes in environmental and socio-economic issues. The situation of geography stems from the fact that in most cases, it is one of the ancillary partner of teams preparing the commissioned documents or reports. The cognitive and methodical aspects, presented in the earlier part of the conclusions, should constitute a base for the preparation of expert reports for various institutions handling social issues. The exceptionality of geographic sciences, compared to other scientific fields, boils down to, among others, the overview of the examined reality geographers can have, both from a technicist and a humanist point of view. Defining relations between science and practice for both of the aforementioned perspectives constitutes a research challenge. It was this issue that the two subsequent application goals of the dissertation referred to. The creative attitude may be defined as something innovative and useful. In theoretical terms, the solutions proposed herein may be described as creative thanks to the indication of singular features, of certain attributes in the geographical approach. By no means is the geographical perspective mechanically assigned to the existing division of geographical sciences into environmental (physical and geographical) or humanistic (socio-economic).

A technicist and humanist approach has been devised for both divisions of this scientific discipline. Therefore, social geography, which undertakes to study social pathology, keeps on attribute in the form of a humanist overview. And yet, it can also actively explore propositions of a technicist nature, which is desired when there is need for a higher applicability of solutions for the socio-economic environment. The geographical procedure of researching global problems in the contemporary world requires sequential order of solutions implemented, although when it is necessary to eliminate one of the proposed measures this does not change the validity of a shortened procedure.

Description and analysis of the selected problems in the contemporary world is an important step in understanding the complexity of today's world. The adopted Central European perspective of looking at the world is devoid of a Eurecentric vision of the world. The authors' great sensitivity and awareness of the complexity of the contemporary world results from the knowledge and experience gained during research and case studies carried out on different continents.

It is also an element of a global discussion in which many entities join and strive to supplement the current discourse. Selected examples expand the base of observations with new spaces and new approaches to reality. They are a creative contribution to the interpretation of ongoing processes, the scale and diversity of which continue to surprise. This book is of interest to a wide audience, aware of the dynamics of changes taking place in the contemporary world and reporting a need for reliable sources of explanations of the mechanisms taking place in various spheres of human existence and activity. The potential recipients of this book are researchers and commentators of global events, journalists, students of various academic professions and everyone interested in understanding the phenomenon of the complexity of the modern globe.

Not all problems of the contemporary world have been explained here, but the indicated interpretive key carries certain universal measures that can be adapted in other places, in other tasks. We hope that this book has given the reader at least partial answers regarding the cognitively important issues facing the world.

Uncited References

10 największych problemów współczesnego świata zdaniem millenialsów (2018) [10 biggest problems of the modern world according to millennials, 2018], 27 lutego 2018, Business Insider Polska (www.businessinsider.com.pl/rozwoj-osobisty/millenialsi-czego-sie-obawiaja-ankieta-global-shapers/8p6zb3p)

Baranowska-Prokop E (2014) Ambush marketing—kontrowersyjne narzędzie kreowania wizerunku marki [Ambush marketing—a controversial tool of the brand image creation. Globalization. Liberalization. Ethics. Zeszyty Naukowe Uniwersytetu Szczecińskiego. Globalizacja. Liberalizacja. Etyka 800(8):192–202

Białek M (2019) Typology of tax havens in the world—Typologia rajów podatkowych na świecie—master thesis. Univeristy of Warsaw, Warsaw, Faculty of Geography and Regional Studies

Britannica. Encyklopaedia Britannica, available from www.britannica.com (20.10.2018)

Castles S (2010) Understanding global migration: a social transformation perspective. J Ethnic Migr Stud 36(10). Theories of Migration and Social change. https://doi.org/10.1080/1369183X.2010.489381

Czarnecka D (2015) Pomniki wdzięczności Armii Czerwonej w Polsce Ludowej i w III Rzeczypospolitej. Instytut Pamięci Narodowej. Komisja Ścigania Zbrodni Przeciwko Narodowi Polskiemu (The Institute of National Remembrance—Commission for the Prosecution of Crimes against the Polish Nation)

Czepczyński M (2010) Interpreting post-socialist icons: from pride and hate towards disappearance and/or assimilation. Hum Geographies 4(1):67–78

Dharmapala D, Hines JR (2006) Which countries become tax havens? Science Direct NBER

Dharmapala D, Hines JR (2009) Which countries become tax havens? J Public Econ 93(9–10):1058–1068

Dimant E, Krieger T, Meierrieks D (2013) The effect of corruption on migration 1985–2000. Appl Econ Lett 20(13):1270–1274. https://doi.org/10.1080/13504851.2013.806776

Engineering. In: Proceedings 13th international conference on wind engineering, Amsterdam

European Commission (2017) Flash Eurobarometr 457. Report. Businesses attitudes towards corruption in the EU. December. TNS Political & Social. European Commission

Eurostat Database, European Commision, https://ec.europa.eu/eurostat/en/web/main/data/database (05.05.2021)

Eurostat Statistics Explained, European Commision https://ec.europa.eu/eurostat/statistics-explained/pdfscache/34962.pdf (05.05.2021)

Farré L, Libertad González, Ortega F (2011) Immigration, family responsibilities and the labour supply of skilled native women. J Econ Anal Policy 11(1)

K. Podhorodecka and T. Wites (eds.), *Global Challenges*, Springer Geography,
https://doi.org/10.1007/978-3-031-60238-2

Flash Eurobarometr 457. Report. Businesses' attitudes towards corruption in the EU. December 2017. TNS Political and Social
Florczak F (2021) Development of sustainable tourism and examples of applying its principles in the area of the Alps—[Rozwój turystyki zrównoważonej oraz przykłady stosowania jej zasad na obszarze Alp]—bachelor degree thesis. Faculty of Geography and Regional Studies. University of Warsaw, Warsaw
Foxnews. Available from www.foxnews.com (19.09.2018)
Gamboni D (2013) The destruction of art: iconoclasm and vandalism since the French Revolution. Reaktion Books
Geoethics at the heart of all geoscience. Ann Geophys 60, Fast Track 7. Available from www.geoethics.org (5.07.2018)
Geoethics. Definition of Geoethics. Available from www.geoethics.org/geoethics (5.07.2018)
Glanz MH, Zonn IS (2005) The Aral sea: water, climate, and environmental change in Central Asia. The World Meteorological Organization. Weather—Climate. Water. WMO-No.982. Geneva
Green P (2005) Disaster by sesign. Coruption, construction and catastrphe. Brit J Criminol 45:528–546
Gułag.info.pl available from http://gulag.info.pl (20.09.2018)
Hines JR (2004) Do tax havens flourish? NBER Working Papers 10936
Human Development Indices and Indicators (2018) Statistical update. United Nations Development Programme, New York
IAPF, International Association for Promoting Geoethics. http://www.geoethics.org/geoethics
Internet Encyclopedia of Philosophy, IEP. Available from www.iep.utm.edu (8.10.2017)
International Association for Promoting Geoethics. IAPG. Definition of geoethics. Available from www.geoethics.org/geoethics (5.07.2018)
Jasiński M (2018) Nasycenie gospodarki turystyką a poziom rozwoju społeczno-ekonomicznego w małych rozwijających się państwach wyspiarskich. In: Zioło Z, Rachwał T (eds), Prace Komisji Geografii Przemysłu Polskiego Towarzystwa Geograficznego. Uwarunkowania rozwoju przedsiębiorstw oraz wybranych sektorów gospodarki, Wydawnictwo Naukowe UP w Krakowie, Warszawa-Kraków
Johnson NC (1995) Cast in stone: monuments, geography, and nationalism. Environ Plann D Soc Space 13(1):51–65
Jones M, Jones R, Woods M, Whitehead M, Dixon D, Hannah M (2014) An introduction to political geography: space, place and politics. Routledge
Jones P (2007) 'Idols in stone' or empty pedestals? Debating revolutionary iconoclasm in the post-Soviet transition. In: Boldrick S, Clay R (eds) Iconoclasm. Contested objects, contested terms. Routledge, pp 241–260
Kantowicz E (1988) Relations between man and environment in research and teaching of regional geography. Miscellanea Geographica 3:181–186
Keller EA, De Vecchio DE (2011) Natural hazards; earth's processes as hazards, disasters, and catastrophes, 3rd edn. Routledge, Pearson Prentice Hall
Komunikat Komisji Europejskiej—Skoordynowana reakcja gospodarcza na epidemię COVID-19—A coordinated economic response to the COVID-19 epidemic—www.eur-lex.europa.eu/legal-content/PL/TXT/?uri=CELEX%3A52020DC0112 (25.12.2020)
Konceptualizacja i dyskusja, (ed) Adam Bochentyn, Jakub H. Szlachetko, Cyfrowa czy analogowa? Funkcjonowanie administracji publicznej w stanie kryzysu. Instytut Metropolitalny. Wyd. Uniwersytetu Gdańskiego (w druku); ISBN 978-83-62198-27-6; ISBN 978-83-62198-28-3 (on-line)
Kostrowicki J (1973) Zarys geografii rolnictwa. PWN, Warszawa
Kuchciak I (2008) Wykorzystanie rajów podatkowych w celu zwiększenia wartości przedsiębiorstwa. Studia i Prace Wydziału Nauk Ekonomicznych i Zarządzania 6:349–361
Kurcil-Białecka R, Raj na ziemi, 'Manager's Life'. www.kancelaria-skarbiec.pl/pdf/managers-life.pdf (27.07.2018)
Lewis J, Kelman I (2012) The good, the bad and the ugly: Disaster Risk Reduction (DRR)

Lewis J (2017) Social impacts of corruption upon community resilience and poverty. J Disaster Risk Studies 9(1), AOSIS
Main I (2004) President of Poland or 'Stalin's most faithful pupil'? The cult of Bolesław Bierut in Stalinist Poland. In: Apor B, Behrends JC, Jones P, Rees EA (eds) The leader cult in communist dictatorships. Palgrave Macmillan, London, pp 179–193
Marples DR (2016) Decommunization, memory laws, and "builders of Ukraine in the 20th century." Acta Slavica Iaponica 39:1–22
Ministerstwo Spraw Zagranicznych. Święta Narodowe Państw Obcych. www.msz.gov.pl/pl/ministerstwo/protokol_dyplomatyczny/swieta_narodowe_panstw_obcych1 (06.02.18)
Nations Unies Annuaire Demographique (2018) United Nations demographic yearbook, department of economic and social affairs. New York
Nijland H (2007) Room for the rivers programme. Cost of flood protection measures in the Netherlands Paris: Programme Directorate Room for the River, Ministry of Transport, Public Works and Water Management. Available at: www.riob.org (6.09.2012)
NIK o inwestycjach powodziowych. NIK Report. https://www.nik.gov.pl/aktualnosci/srodowisko/nik-o-planowaniu-i-realizacji-inwestycji-na-terenach-powodziowych.html (20.10.2015)
OECD (1998) Harmful Tax Competition: An Emerging Global Issue, zwany Raportem o szkodliwej konkurencji podatkowej. Paryż, www.uniset.ca/microstates/oecd_44430243.pdf (06.08.2018)
OECD (2000) Towards global tax co-operation: progress in identifying and eliminating harmful tax practices. Paryż, www.oecd.org/tax/transparency/about-the-global-forum/publications/towards-global-tax-cooperation-progress.pdf (09.08.2018)
Oxford Dictionary of Philosophy. www.oxfordreference.com/view/10.1093/acref/9780199541430.001.0001/acref-9780199541430
Passports for sale. 60 Minutes (2017) CBS 1 January, New York, Columbia Broadcasting System, Daily Observer (Antigua and Barbuda)
Poleszczuk G (2000) Raje podatkowe w ujęciu polskiego systemu prawa podatkowego. Częstochowskie Wydawnictwo Naukowe przy Wyższej Szkole Zarządzania, Częstochowa
Polner M. Ireland R., WCO, World Custom Organization
Prieur M (2012) Ethical Principles on Disaster Risk Reduction and People's Resilience. EUR-OPA. https://www.coe.int/t/dg4/majorhazards/ressources/pub/Ethical-Principles-Publication_EN.pdf (5.09.2022)
Red List (2000) ICUN Red list categories and criteria version 3.1, 2nd edn. Prepared by the IUCN Species Survival Commission, IUCN CouncilGland, Switzerland
Regions in Europe—Statistics visualised. European Commision. https://ec.europa.eu/eurostat/cache/digpub/regions (05.05.2021)
Rozporządzenie Ministra Finansów w sprawie określenia krajów i terytoriów stosujących szkodliwą konkurencję podatkową dla celów podatku dochodowego od osób prawnych: z dnia 11 grudnia 2000 r., 16 maja 2005 r., 9 kwietnia 2013 r., 23 kwietnia 2015 r., 17 maja 2017 r
Rucińska D (2015) Spatial distribution of flood risk and quality of spatial management: case study in Odra valley, Poland. Risk Anal 35(2):241–251
Rucińska D, Walewski A (2005) Kształcenie w Zakładzie Geografii Regionalnej. Prace i Studia Geograficzne, 34, pp 57–67
Ruskulis O (2002) Developing processes for improving disaster mitigation of the urban poor Basin News (Building Advisory Service and Information Network), No 23 June. SKAT St Gallen
Sabur B (2015) Effect of corruption on natural disasters vulnerability. Submitted to central European university school of public policy. In: Partial fulfillment of the requirements for the degree of Master of Public Policy (MPA), Supervisor: Cristina Corduneanu-Huci, Budapest
Seasonality in tourism demand. Eurostat Statistics Explained. https://ec.europa.eu/eurostat/statistics-explained/index.php?title=Seasonality_in_tourism_demand#Nearly_one_in_four_trips_of_EU_residents_made_in_July_and_August (05.05.2021)
Sendai Framework for Disaster Risk Reduction 2015–2030. UNDRR. www.undrr.org/publication/sendai-framework-disaster-risk-reduction-2015-2030

SERC. Science Education Resource Center at Carleton College. Teaching geoethics across the curriculum. What is GeoEthics? Available from www.serc.carleton.edu (5.07.2018)

Startribune. www.startribune.com/politics/national/232508921.html (19.11.2013)

The human cost of disasters 2000–2019. www.reliefweb.int/report/world/human-cost-disasters-overview-last-20-years-2000-2019 (30.01.2021)

The nature concervancy. The threat of invasive species disrupting the natural balance. http://www.nature.org/ourinitiatives/habitats/forests/explore/the-threat-of-invasive-species.xml (10.07.2021)

Tourism 2020 Vision. Global Forecats (2001) World tourism organization, vol 7, Madrid

Tourism in numbers. Slovenian Tourist Board. https://www.slovenia.info/en/business/research-and-analysis/tourism-in-numbers (05.05.2021)

Transparency International, Corruption Perceptions Index (CPI)

Turystyka. Dokumenty informacyjne o Unii Europejskiej—2021. https://www.europarl.europa.eu/ftu/pdf/pl/FTU_3.4.12.pdf (05.05.2021)

Turystyka.wp.pl Droga Umarłych www.turystyka.wp.pl (18.09.2018)

UNHCR (2015) 2015: The year of Europe's refugee crisis. www.unhcr.org/news/stories/2015/12/56ec1ebde/2015-year-europes-refugee-crisis.html

Unia Europejska przekaże 430 mln euro pomocy i żandarmerię. www.wiadomosci.wp.pl/kat,1356,title,UE-przekaze-Haiti-430-mln-euro-pomocy-i-zandarmerie,wid,11868869,wiadomosc.html?ticaid=1161e0 (10.12.2015)

Urzędowy wykaz nazw państw i terytoriów niesamodzielnych (2013) Komisja Standaryzacji Nazw Geograficznych poza Granicami Rzeczypospolitej Polskiej (KSNG), Główny Urząd Geodezji i Kartografii, Warszawa

Urzędowy wykaz nazw państw i terytoriów niesamodzielnych (2017) Komisja Standaryzacji Nazw Geograficznych poza Granicami Rzeczypospolitej Polskiej (KSNG), Główny Urząd Geodezji i Kartografii, Warszawa

Usbr.gov Reclamation, Managing Water in the West. Hoover Dam. The Story of Hoover Dam—Essays. Fatalities at Hoover Dam. Available from www.usbr.gov (17.09.2018)

Ustawa z dnia 29 sierpnia 1997 r. Ordynacja podatkowa. Kancelaria Sejmu

Van Cant K (2009) Historical memory in post-communist Poland: Warsaw's Monuments after 1989. Stud Slavic Cult 8:90–119

Versus Disaster Risk Creation (DRC) Perspective

Vertovec S (2007) Super0diversity and its Implications. Ethnic Racial Stud 30(6):1024–1054 20(4):699–751

Volontourism. www.worldvision.ca/stories/voluntourism-the-good-and-the-bad (10.12.2020)

Wojtczuk M (2015) Rada Warszawy: Czterej Śpiący już nie wrócą [Warsaw Council: Czterech Śpiących will not return], Gazeta Wyborcza. Available from: https://warszawa.wyborcza.pl/warszawa/1,34862,17486935,Rada_Warszawy___Czterej_Spiacy__juz_nie_wroca.html (22.12.2021)

Wood N, Jones J, Schelling J, Schmidtlein M (2014) Tsunami vertical-evacuation planning in the U.S. Pacific Northwest as a geospatial, multi-criteria decision problem. Int J Disaster Risk Reduction 9:68–83

World Bank. World Bank to Provide an Additional $100 Million to Haiti, Following Earthquake. http://web.worldbank.org/WBSITE/EXTERNAL/NEWS/0,,contentMDK:22440632~pagePK:64257043~piPK:437376~theSitePK:4607,00.html (13.12.2015)

Wspólnota Narodów [w:] Encyklopedia PWN (internetowa), Wydawnictwo Naukowe PWN, Warszawa

www.blog.nationalgeographic.org/2019/09/25/the-global-impacts-of-habitat-destruction (11.05.2021)

www.britannica.com/topic/moral-hazard (28.09.2018)

www.businessinsider.com.pl/firmy/wielkie-statki-wycieczkowe-trzeba-zlomowac-bo-zatopil-je-kryzys-branza-walczy-o/3nxvfeb

www.businessinsider.com.pl/firmy/wplyw-pandemii-koronawirusa-na-rynek-pracy/yz4bcsq (25.12.2020)

www.ceo.com.pl/paszport-na-sprzedaz-najlepsze-obywatelstwa-ekonomiczne-43925
www.cia.gov/library/publications/the-world-factbook/index.html—CIA, The World Factbook 2017 (20.03.2018)
www.clustercollaboration.eu/content/commission-welcomes-approval-recovery-and-resilience-facility (13.04.2021)
www.data.worldbank.org—Bank Światowy (20.03.2018)
www.data.worldbank.org World Bank (02.2022)
www.data.worldbank.org (17.05.2021)
www.data.worldbank.org/indicator—World Bank (25.12.2020)
www.ec.europa.eu/info/business-economy-euro/recovery-coronavirus/recovery-and-resilience-fac ility_en (13.04.2021)
www.ec.europa.eu/info/live-work-travel-eu/coronavirus-response/jobs-and-economy-during-cor onavirus-pandemic_pl#prognozy-gospodarcze—European Comission (25.12.2020)
www.elephantconservation.org (12.07.2021)
www.encyklopedia.pwn.pl/haslo/Wspolnota-Narodow;3998422.html (28.05.2019)
www.forbes.pl/gospodarka/epidemia-koronawirusa-i-jej-wplyw-na-gospodarke-globalne-spowol nienie-moze-byc/m8p0chc—Forbes
www.gov.pl/web/rozwoj-praca-technologia (09.04.2021)
www.iop.krakow.pl/pckz/default2b15.html?nazwa=katz&je=pl (26.04.2021)
www.ipbes.net (24.04.2021)
www.islands.unep.ch/Tisolat.htm (26.04.2021)
www.klimada.mos.gov.pl/blog/2013/04/15/roznorodnosc-biologiczna (26.04.2021)
www.ksng.gugik.gov.pl/pliki/zmiany_nazw_panstw.pdf—Zmiany w nazewnictwie państw i terytoriów niesamodzielnych, 2018, Komisja Standaryzacji Nazw Geograficznych poza Granicami Rzeczypospolitej Polskiej (KSNG) (06.02.2019)
www.naukawpolsce.pap.pl/aktualnosci/news%2C76959%2Craport-milion-gatunkow-zagroz onych-wyginieciem.html (24.04.2021)
www.naukawpolsce.pap.pl/aktualnosci/news%2C80107%2Czmiany-klimatyczne-nie-jedyne-zag rozenie-dla-gatunkow.html (24.04.2021)
www.newsweek.pl/wiedza/szoste-wymieranie-wstrzasajacy-raport-onz-o-przyszlosci-naszej-pla nety/x3552n2 (24.04.2021)
www.polityka.pl/tygodnikpolityka/rynek/1772622,1,obywatelstwo-na-sprzedaz.read
www.prawo.uni.wroc.pl—Globalne problemy współczesnego świata (8.08.2022)
www.pzuzdrowie.pl/poradnik-o-zdrowiu/koronawirus-covid-19/najwieksze-pandemie-swiata (09.05.2021)
www.rhinos.org (01.05.2021)
www.savetherhino.org (05.2021)
www.theguardian.com/world/2018/jun/02/citizenship-by-investment-passport-super-rich-nation ality (07.06.2018)
www.transparency.org/whatwedo/nis (12.02.2020)
www.unhcr.org/refugee-statistics—UNHCR (02.2022)
www.unwto.org (16.01.2020)
www.unwto.org (17.05.2016)
www.wfos.gdansk.pl/wiadomosci/bioroznorodnosc-dlaczego-tak-wiele-od-niej-zalezy (24.04.2021)
www.wiadomosci.gazeta.pl/wiadomosci/1,114871,12170988,21_bilionow_dolarow__Tyle_swia towa_elita_bogaczy_ukryla.html
www.wprost.pl/swiat/10446001/szklany-most-w-chinach-rozpadl-sie-przez-wiatr-mezczyzna-wal czyl-o-zycie.html (11.05.2021)
www.zielonewiadomosci.pl/tematy/ekologia/dlaczego-utrata-bioroznorodnosci-szkodzi-ludziom-tak-samo-jak-zmiany-klimatu (24.04.2021)
Zrównoważony rozwój, Ministerstwo Rozwoju, Pracy i Technologii. https://www.gov.pl/web/roz woj-praca-technologia/zrownowazony-rozwoj (05.05.2021)

Index

K. Podhorodecka and T. Wites (eds.), *Global Challenges*, Springer Geography,
https://doi.org/10.1007/978-3-031-60238-2